The Road to OPEC

TEXAS PAN AMERICAN SERIES

The Road to OPEC

UNITED STATES RELATIONS WITH VENEZUELA, 1919–1976

by Stephen G. Rabe

UNIVERSITY OF TEXAS PRESS, AUSTIN

FOR MY MOTHER AND FATHER

The Texas Pan American Series is published with the assistance of a revolving publication fund established by the Pan American Sulphur Company.

Publication of this book was also assisted by a grant from the University of Texas at Dallas.

Copyright © 1982 by the University of Texas Press
All rights reserved
Printed in the United States of America
First Edition, 1982

Requests for permission to reproduce material from this work should be sent to Permissions, University of Texas Press, Box 7819, Austin, Texas 78712.

LIBRARY OF CONGRESS CATALOGING IN PUBLICATION DATA
Rabe, Stephen G.
The road to OPEC.
(Texas Pan American series)
Bibliography: p.
Includes index.
1. Petroleum industry and trade—Venezuela—History. 2. Petroleum industry and trade—United States—History. 3. United States—Foreign economic relations—Venezuela. 4. Venezuela—Foreign economic relations—United States. I. Title.
HD9574.V42R3 337.73087 81-16000
ISBN 0-292-76020-5 AACR2

Contents

831995

LIBRARY
ALMA COLLEGE
ALMA, MICHIGAN

Preface

The United States and Venezuela have interacted dynamically in the twentieth century. Venezuela has been a prime recipient of U.S. capital, a significant trading partner, and a testing ground for U.S. development and reform programs. The exploitation of Venezuela's oilfields, principally by U.S. oil companies, and the ensuing surge in income have profoundly altered Venezuela's political and economic life. On the other hand, the political security and economic welfare of the United States have come to depend upon petroleum-exporting nations, such as Venezuela.

This study focuses on the period between 1919 and 1976, the years when U.S. oil companies operated the Venezuelan oil industry. Until 1919, U.S. diplomats and businessmen attached no special significance to relations with Venezuela, a weak agricultural nation with seemingly minor economic potential. The experience of World War I radically changed this perception of Venezuela's value. Fueling Allied war machines drained U.S. oilfields; geologists judged that the United States had only a twenty-year supply of oil reserves. They became further alarmed when they discovered that British oil companies, with the active support of their government, had captured the world's most promising oilfields, including Venezuela's. In the 1920s, the Department of State responded to the postwar energy crisis with a concerted effort to loosen the British hold on Venezuelan oil and open the area to U.S. capital. Persistent U.S. diplomatic pressure, a large investment by U.S. oil companies, and unusual success in drilling combined to establish U.S. dominance in Venezuela's oilfields by 1929. Indeed, so successful were oilmen that they turned Venezuela into the world's leading exporter of oil, a position the nation held until 1970.

By the mid-1920s, geologists were finding the vast new oilfields in the United States that would supply its energy needs for almost fifty years. But the United States continued to stress its relations with Venezuela. Oilmen steadily expanded their holdings there, giving Venezuela the largest U.S. direct investment in Latin America, and the financial returns on this investment were handsome. U.S. exporters also counted Venezuela among their ten best customers. Beyond wishing to protect and promote a lucrative market, the State Department cultivated relations with Venezuela because it wanted access to the South American nation's vital raw material during emergencies like World War II, the Korean War, and the Arab-Israeli wars.

Not only are significant bilateral issues examined here but also Venezuelan-U.S. relations are placed within the context of the Latin American policy of the United States. Key policies and programs, such as the Open Door, the Good Neighbor, reciprocal trade agreements, Export-Import Bank loans, and the Alliance for Progress, were evident in Venezuela. The study analyzes these initiatives and debates interpretations that scholars have assigned to them.

In addition to serving as a case study of inter-American relations, this investigation provides insight into both the politics of the contemporary energy crisis and the growing split between raw-material producers and their industrial customers, the so-called North-South debate. The two nations continually bickered about the efficacy of the liberal trade and investment principles of the United States for Venezuela. While conceding that they initially lacked the resources and expertise to produce oil, Venezuelan officials gradually opposed U.S. policies by curtailing the power and prerogatives of the foreign oil industry. They also argued that the law of supply and demand, the pricing mechanism interwoven with the principles of free trade and private foreign investment, could not set fairly the price of oil, an exhaustible natural resource essential to modern civilization. Venezuela founded in 1960 the Organization of Petroleum Exporting Countries (OPEC), nurtured the cartel through the 1960s, joined with other oil exporters in quadrupling prices in late 1973, and on January 1, 1976, became the first major producer since Mexico to nationalize U.S.-owned properties.

A final task of the study is to interpret how the United States influenced Venezuela's development. During the twentieth century, Venezuela has evolved remarkably from a poor, agrarian society with a turbulent political history into an urban nation with the highest per capita income in Latin America and a stable, representative government. The capital, technology, ideologies, and life-styles that U.S. diplomats and businessmen brought to Venezuela affected both the patterns and the pace of Venezuela's modernization.

In investigating and writing this book, I incurred many scholarly debts. I should like to thank the staffs of the National Archives and the Library of Congress in Washington, D.C., the Public Record Office in London, the Nettie Lee Benson Library at the University of Texas at Austin, and the Biblioteca Nacional in Caracas. I invariably found the archivists at the Hoover, Roosevelt, Truman, Eisenhower, Kennedy, and Johnson presidential libraries to be both efficient and helpful. In particular, the archivists at the Kennedy and Johnson libraries assisted me in using the Freedom of Information Act to declassify documents. I also gratefully acknowledge the help that I received from the staffs who guided me through manuscript collections at the following schools: Clemson University, Columbia University, Harvard University, Princeton University, University of Vermont, Yale University, and Kings College in Aberdeen, Scotland. Without financial aid, it would have been impossible to visit these institutions. The Hoover Presidential Library Association, the Lyndon Baines Johnson Foundation, the Harry S Truman Library Institute, the Eleanor Roosevelt Institute, and the University of Connecticut Research Foundation all generously supported my research.

Individual scholars were also kind to me. Professors Hugh M. Hamill, A. William Hoglund, and Thomas G. Paterson of the University of Connecticut painstakingly read this study when it was a doctoral dissertation, and Professor Paterson has continued to be a valued critic of my work. Professor Winfield Burggraaff of the University of Missouri also carefully read my study and provided me with keen insights into the political culture of Venezuela.

Finally, I want to thank my wife, Genice Ann Gladow Rabe, for her encouragement, support, and love.

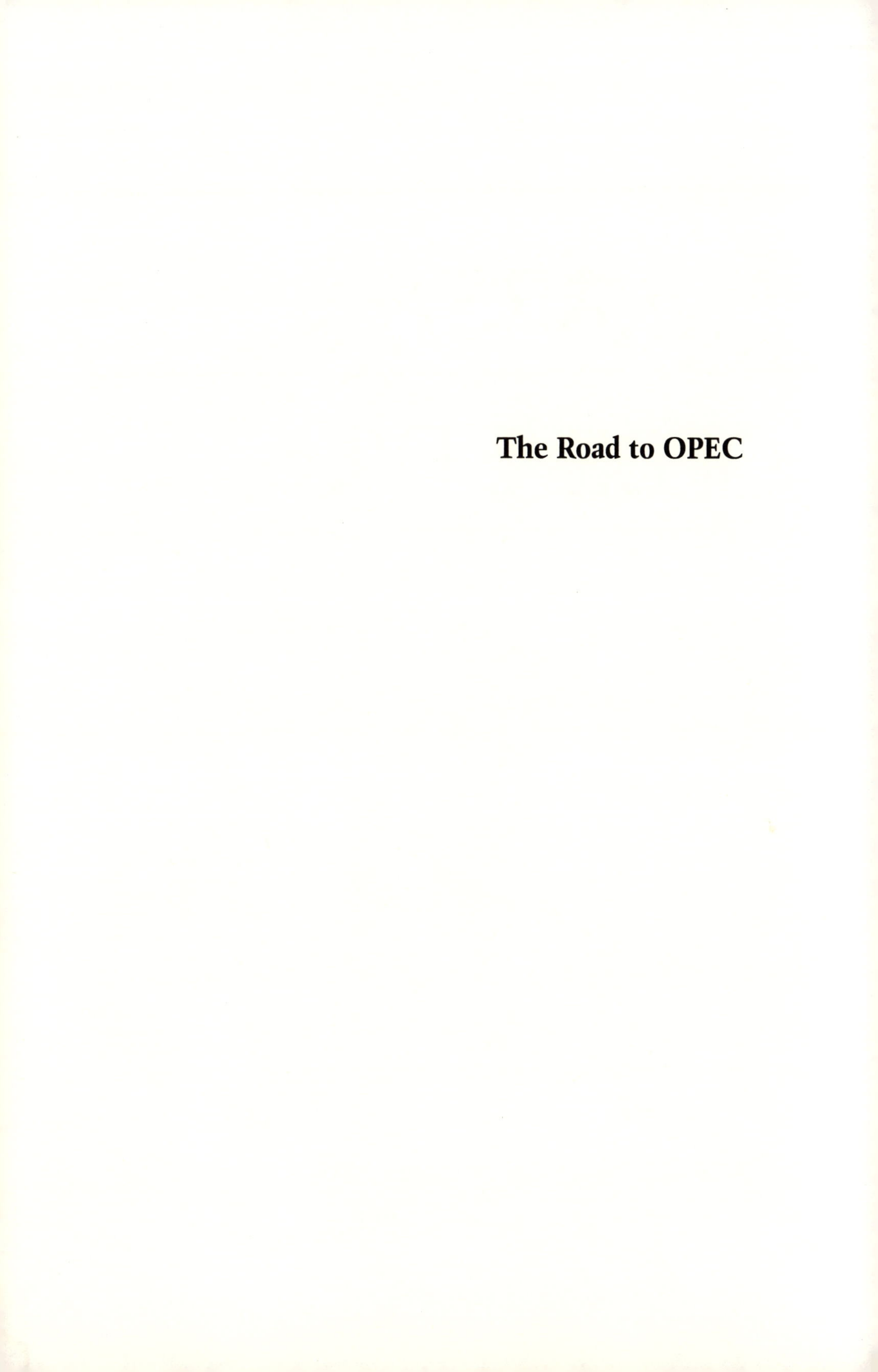

The Road to OPEC

1. The Caribbean Sphere of Influence

Prior to the energy crisis that followed World War I, U.S. officials attached no special significance to diplomatic relations with Venezuela. The nation lacked the traditional attractions a small country might have for an international power. It had neither money nor military strength to aid the United States during an emergency. It had no essential raw materials to exploit or sell. Moreover, it barely contributed to U.S. prosperity. U.S. businessmen conducted less than 2 percent of their Latin American trade with Venezuela, and their investments there were similarly inconsequential. Yet, despite Venezuela's seemingly slight commercial and military value, the United States intervened actively in Venezuelan politics in the late nineteenth and early twentieth centuries. Venezuela formed part of the Caribbean littoral, and events in Venezuela afforded the United States the opportunity to establish further its hegemony in that strategic area. Thereafter, U.S. diplomats labored to keep Venezuela a secure and stable nation within the Caribbean sphere of influence.

An impoverished agricultural nation, Venezuela played only a small role in the world economy in the nineteenth and early twentieth centuries. Its small population, poor agricultural conditions, and incessant domestic unrest together produced the country's backward and stagnant economy. Despite its size—one and one-half times the area of Texas—Venezuela had a minute population. At the time of national independence in 1830, fewer than 1 million people lived in Venezuela, and, seventy years later, the population had increased to only 2.4 million people. Warfare and disease, especially dysentery and malaria, retarded population growth.[1]

The vast majority of Venezuelans eked out a living as subsistence farmers. The typical Venezuelan peasant of the early twentieth century raised corn, beans, and bananas for his own family's consumption. He was rarely a proprietor, for, as was typical throughout Latin America, less than 10 percent of the population, a privileged elite, controlled the choice areas. Potentially fertile land often remained fallow. And the peasant had little opportunity to expand his output because Venezuela lacked an abundance of arable land. Forced to cultivate small plots, the average farmer suffered low output. Prior to World War I, Venezuela could therefore barely produce enough food for the needs of its small population.[2]

Continual domestic strife compounded Venezuela's economic problems. Except for the tranquil period from 1830 to 1848 during the rule of the so-called Conservative Oligarchy, military conflict ravaged Venezuela between 1810 and 1909. The struggle for independence from Spain (1810–1821), for example, cost Venezuela one-fourth of its population. The bloody Federal War of 1858 to 1864 damaged the country's economy even more than had the independence movement. Cattlemen had profitably raised, since colonial times, livestock on the broad *llanos*, or plains, of central Venezuela. By 1858, 12 million head of cattle roamed through the central states of Apure and Guárico. The effects of the Federal War reduced the number of cattle in half by 1864 and so weakened the cattle industry that only 1.5 million animals existed by 1910. Moreover, the Federal War destroyed the authority and power of the central government and turned Venezuela into a nation of competing, hostile regions under the control of local strong men, or *caudillos*. The man on horseback dominated Venezuelan life. Of the 184 members of the legislature in the mid-1890s, 112 of them claimed the rank of general.[3] National economic progress was impossible in such a violent political system.

Coffee production was the only dynamic sector of the Venezuelan economy. In the late nineteenth century, production expanded in the temperate zones of the west Andean states. By 1914, Venezuela harvested over one million bags per year and exported more coffee than any other country except Brazil. As the exports of coffee increased, however, revenues from two traditional exports—cacao and cattlehides—declined. Competition from other countries limited the cacao market, and the dearth of animals severly hampered Venezuela's ability to sell hides. In addition, the coffee boom had uneven effects on the economy. By absorbing native capital, coffee production denied credit to other sectors of the economy and left the nation dependent on foreign merchants, particularly Germans, for loans. Since they had a stake in preserving traditional patterns of trade, these merchants successfully opposed, for example, tariffs which might diversify the economy by nurturing local industries.[4]

Prior to World War I, then, few observers could have foreseen that Venezuela would become commercially important to the world. To be sure, Venezuelans knew that their country had petroleum deposits. Natural seepages of petroleum had been occurring around the Lake Maracaibo region for centuries. The earliest Spanish explorers, for example, observed Indians using the sticky substance to caulk and repair their canoes. Widespread petroleum use in the United States from the 1860s on encouraged some Venezuelans to consider the commercial possibilities of this natural resource. Several states granted land concessions to oil speculators, and in 1879, Manuel Antonio Pulido's Compañía Petrolera del Táchira began production in the Rubio district in the Venezuelan Andes, close to the Colombian border. Such operations were limited, isolated, and primitive; Pulido's company produced only forty gallons of petroleum a day by digging pits and actually scooping the petroleum out with buckets.[5] Venezuela lacked both the capital and the technology to exploit its petroleum deposits further, and no other nation evinced much interest in those natural petroleum wells. It would take the results of the Venezuelan political upheaval of December 19, 1908, and the conclusion of the world war in November 1918 before foreigners exploited Venezuela's petroleum potential.

In view of Venezuela's poverty and the U.S. preoccupation with continental expansion and the problem of slavery, relations between the two young nations were, during most of the nineteenth century, predictably routine and internationally insignificant. Questions of expatriation and naturalization, political asylum and diplomatic immunity, and free commerce and neutral rights were the stuff of U.S.-Venezuelan relations.[6] Prior to the 1890s, commercial exchanges were also limited and infrequent. U.S. direct investments in Venezuela probably amounted to less than $1 million. As in the rest of South America, European investments in Venezuela far outranked those of the United States. At the beginning of the twentieth century, the British minister in Caracas estimated British investments at $60 million. British capital financed major railways, such as the lines between the port of La Guaira and Caracas and between Puerto Cabello and Valencia, and controlled the country's public utilities. German investments of perhaps $40 million were located in railroads and merchant houses, with one thousand German merchants residing in Caracas. While larger than U.S. investments, the amount of European capital in Venezuela was small compared, for example, to the more than $1 billion that British entrepreneurs invested in Argentina.[7]

Europeans also dominated Venezuela's trade, providing it with at least 75 percent of its imports during the nineteenth century. The

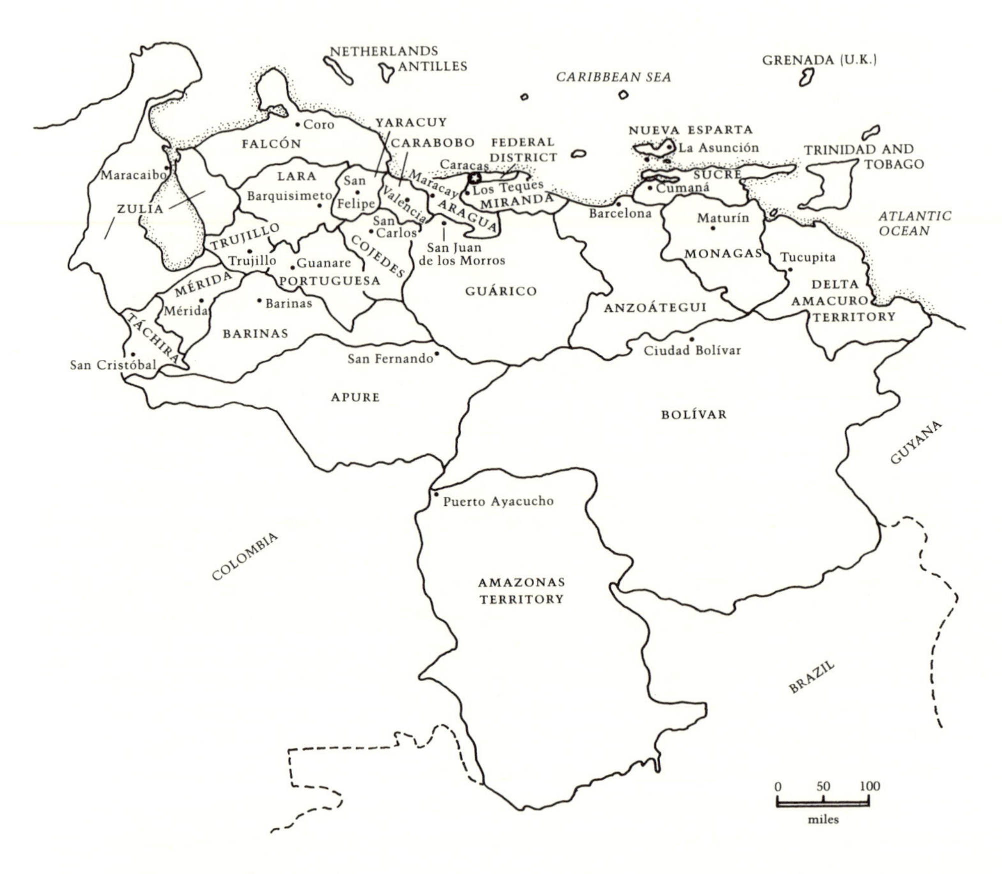
NETHERLANDS
ANTILLES
CARIBBEAN SEA
GRENADA (U.K.)
Coro
YARACUY
FALCÓN
CARABOBO
FEDERAL
DISTRICT
NUEVA ESPARTA
La Asunción
TRINIDAD AND
TOBAGO
Maracaibo
LARA
Caracas
Los Teques
SUCRE
Cumaná
San
Felipe
Barquisimeto
Valencia
Maracay
ARAGUA
MIRANDA
ZULIA
Barcelona
Maturín
ATLANTIC
OCEAN
TRUJILLO
San
Carlos
Trujillo
Guanare
COJEDES
San Juan
de los Morros
MONAGAS
Tucupita
MÉRIDA
PORTUGUESA
GUÁRICO
DELTA
AMACURO
TERRITORY
Mérida
Barinas
ANZOÁTEGUI
TÁCHIRA
BARINAS
Ciudad Bolívar
San Cristóbal
San Fernando
APURE
BOLÍVAR
GUYANA
Puerto Ayacucho
COLOMBIA
AMAZONAS
TERRITORY
BRAZIL
0
50
100
miles

Venezuela

British captured the largest share of the Venezuelan market, for they reaped the advantages of being the leader in the industrial revolution and of having been the first power to obtain a commercial treaty with Venezuela. As such, British businessmen readily sold their textiles and machinery in Venezuela, albeit their sales were small compared to their lucrative trade in Argentina, Brazil, and Chile. U.S. exporters could not compete during the nineteenth century with their European counterparts. The "infant industries" of the United States were not yet as efficient as British factories, and Venezuela imposed high tariffs on agricultural imports. U.S. sales in Venezuela were unimpressive, about $3 million a year, for example, during the 1880s. Indeed, the balance of trade favored Venezuela. By marketing about half of its coffee production in the United States, Venezuelan traders earned approximately $6.5 million a year in the 1880s.[8]

While, during most of the nineteenth century, the presence of the United States in Venezuela was limited, the North American nation was an emerging industrial giant with international ambitions. These aspirations would be revealed in the first noteworthy diplomatic event in Venezuelan-U.S. relations, the Venezuelan–British Guianan boundary dispute. For over fifty years, Great Britain and Venezuela had haggled over the exact location of the boundary separating Venezuela and the British colony of Guiana. In the 1870s and 1880s, Venezuela implored the United States to support its boundary claim. The United States made, however, only mild representations to the British. In 1895 U.S. officials decided to enter the dispute. In a stridently worded note of July 20 to the Foreign Office, President Grover Cleveland and Secretary of State Richard Olney demanded that Great Britain state whether it would submit the disputed territory issue to arbitration. The two powers heatedly debated the issue for several months, but, in the end, Britain agreed to arbitration. The British concluded that it was foolish to contest U.S. power over a few thousand square miles of land, and, in any case, it had enough difficulties with the Boer War.[9]

Though ostensibly aiding Venezuela, the United States ignored that nation's desires. Cleveland and Olney never consulted Venezuelan officials about their July 20 protest, and, when Britain submitted, they excluded Venezuela from the arbitral commission. After vehement protests, including street rioting in Caracas, Venezuela gained the right to appoint one arbitrator, but not a Venezuelan. Nonetheless, the United States and Great Britain had agreed beforehand to exempt from arbitration those areas settled for more than fifty years by British subjects. These were the very areas that the Venezuelans had hoped to secure from arbitration.[10]

The Cleveland administration used the boundary dispute to achieve two goals. First, it gained for Venezuela control of the mouth of the Orinoco River. U.S. officials had feared that, if Britain prevailed in its claim to the Orinoco's mouth, British businessmen would use the river system to dominate the potentially rich commerce of northern South America. That was particularly alarming to President Cleveland, who believed that industrial overproduction had caused the frightening economic depression of the mid-1890s. The United States needed new markets to empty its glutted warehouses and return workers to the factory, and U.S. businessmen agreed with Cleveland's analysis. In 1898, the National Association of Manufacturers set up their first overseas warehouse display of U.S.-manufactured goods in Caracas. U.S. investors, particularly an asphalt company and the Orinoco Company, a Minnesota-based firm which planned to colonize the Orinoco region and exploit its natural resources, also first entered Venezuela in the 1890s. Significantly, the Orinoco's mouth was the only key area in the disputed territory that the United States won for Venezuela.[11] Second, by forcing Britain to accept arbitration, the United States demonstrated its suzerainty over the Caribbean area. U.S. officials were uncertain where the imperialist powers of Europe might encroach next. The Cleveland administration, flexing new industrial muscle, wanted to serve notice that the United States now had the right and responsibility to settle disputes in the Caribbean, or, as Olney argued, "its infinite resources combined with its isolated position render it master of the situation and practically invulnerable as against any or all other powers."[12] The United States had the power to enforce the Monroe Doctrine as never before.

The next diplomatic crisis broke in 1902 and was related to a fundamental change in Venezuelan political life. In 1899, a *caudillo*, General Cipriano Castro of the Andean state of Táchira, led his followers into the capital city of Caracas and seized the central government. Castro, the *mestizo* son of a middle-class merchant, represented the area and men who had prospered from Venezuela's coffee production boom. They wanted political power and spoils commensurate with their economic might. With the help of his efficient field commander, General Juan Vicente Gómez, Castro defeated the various regional *caudillos* in a series of battles and reestablished the authority of the central government in 1902. His triumph brought a semblance of domestic peace to Venezuela and heralded the beginning of more than fifty years of rule by military leaders from Táchira.[13]

Castro did not earn his reputation or notoriety from his exploits in Venezuela, but rather from his cavalier treatment of foreign powers. During the late nineteenth century, the perenially indigent Venezuelan government had borrowed heavily and had sold bonds in Europe. Alleging that foreigners had conspired against his regime, Castro refused to honor the debts, which would eventually amount to over $25 million. In any case, the Venezuelan would have had difficulty repaying, since government revenues declined drastically during his regime because of a depression in the international coffee market. Great Britain, Germany, and Italy responded to Castro's obstinancy by blockading the Venezuelan coast in late 1902. The United States did not initially protest the blockade, for President Theodore Roosevelt did not consider debt collection a violation of the Monroe Doctrine. As he observed to German diplomat Speck von Sternburg in 1901: "If any South American country misbehaves toward any European country, let the European country spank it."[14] During the crisis, as European ships plied Caribbean waters, the president reconsidered his position. He had grown suspicious of German naval maneuvers; he feared that Germany might invade Venezuela and occupy a port. He pressured Kaiser Wilhelm II into lifting the blockade and accepting a U.S.-arranged arbitration of the debt. Since the United States was in the process of securing an isthmian canal, Roosevelt concluded tht he could no longer allow European nations to interfere in Caribbean politics.[15] Roosevelt's experience with the naval blockade coupled with the successful detachment of Panama from Colombia in late 1903 led him to formulate a new policy. In December 1904, in his annual message to Congress, he outlined the "Roosevelt Corollary" to the Monroe Doctrine. To protect its strategic interests, in his view, the United States could properly exercise an "international police power" to ensure that Caribbean nations paid their European creditors.[16]

Though they were diplomatic watersheds, neither the boundary dispute nor the debt controversy specifically involved Venezuelan-U.S. issues. Both were part of the U.S. drive for hegemony in the Caribbean in the years immediately preceding and following the Spanish-American War. But, in order to uphold its claim to full responsibility for Caribbean affairs, the Roosevelt administration tried to maintain order and thereby precipitated the first Venezuelan-U.S. confrontation.

Despite his chastisement by the European powers in 1902–1903, President Castro continued to misbehave. He reneged on his promise to repay promptly Venezuela's international debts, and he harassed foreign diplomats stationed in Caracas. In May 1908, for example, Castro accused the minister from the Netherlands of political intrigue, dismissed him, and interdicted trade between Venezuela and the Dutch West Indies. Moreover, he expropriated U.S. property and annulled U.S. concessions. He posed as a nationalist intent on freeing Venezuela from the clutches of the imperialist powers.[17] As such, he struck a responsive chord among his countrymen, for foreigners had meddled in domestic politics. Venezuelans knew that the New York and Bermúdez Company, the U.S. asphalt company working in their nation, contributed $100,000 to General Manuel A. Matos' attempt in 1901–1902 to overthrow Castro. That bloody rebellion cost the government $5 million and the nation twelve thousand lives.[18] Another U.S. enterprise, the Orinoco Company, beseeched the State Department in 1906 to seize a Venezuelan customs house after Castro's Federal Court annulled its concession. The Orinoco claimed the inviolability of its property rights, but, as the court noted, the company had failed to develop its fifteen-million-acre concession.[19] Castro probably planned to resell the confiscated properties and keep the profits. But, by championing Venezuela's interests against foreign exploitation, the self-styled "Lion of the Andes" undoubtedly made his thievery palatable to the citizenry.

Castro's ridicule of the Western industrial nations infuriated President Roosevelt, who would not tolerate "chronic wrongdoing" in the growing U.S. sphere of influence. Since proclaiming his interpretation of the Monroe Doctrine, Roosevelt had intervened twice in Caribbean nations. In 1905, he persuaded the Dominican Republic to accept U.S. control over its custom houses, and, in late 1906, he reintroduced U.S. troops into Cuba. He also contemplated deploying military force against Castro. When special envoy William J. Calhoun, formerly a U.S. representative in Cuba, failed to convince Castro to honor his international debts, Roosevelt instructed the general staff in 1906 to prepare a campaign plan against Venezuela in sufficient strength "to minimize the chance for effective resistance." The general staff worked from information that military attachés to the legation in Caracas had been gathering since 1903 on the Venezuelan army, its major installations, and the nation's roads and topography.[20] Roosevelt had to shelve the plan, however, for he feared Congress would oppose an invasion. The Democrats in the Senate had already blocked, for example, his customs house treaty with the Dominican Republic.[21] Roosevelt remained belligerent. In 1908 he proposed to Secretary of State Elihu Root that the Marines land on Venezuelan shores and occupy a customs house. As Roosevelt had once remarked, the seizure of a customs house would "show those Dagos that they will have to behave decently." Root shared the president's contempt for Castro, calling him a "crazy brute," but he counseled against war and convinced Roosevelt to submit the Venezuelan matter to Congress, without recommendation, for advice. Once again, Congress declined to act. All Roosevelt could do was to send increasingly harsh notes to the Foreign Ministry and, finally, to sever relations with Venezuela on June 13, 1908.[22]

Six months later, General Gómez solved the Castro problem for Roosevelt. On November 24, 1908, President Castro left Caracas and sailed for Germany to obtain medical treatment for a kidney ailment. The previous day he had turned the government over to his military colleague and vice-president, Gómez. To Castro's dismay, Gómez, alleging that Castro's partisans threatened his life, used his absence to usurp the presidency on December 19. In truth, Gómez was fulfilling his own ambitions and those of his fellow *andinos*, who felt that Castro had slighted them since seizing power.[23] After taking power, Gómez elicited foreign support for his *golpe de estado*. By December 21, he had contacted the various legations and proposed the settlement of outstanding international questions in return for diplomatic recognition and support. The foreign powers responded favorably to Gómez' offer, and within a few months they restored relations with Venezuela. Gómez was then free to consolidate a rule that was to last for twenty-seven years.

Though soliciting the support of all major powers, Gómez sought particularly U.S. aid. On December 21, he requested the call of a U.S. warship to the port of La Guaira "in prevision of events."[24] On the same day, the United States dispatched three warships to the Venezuelan coast. One, the battleship *North Carolina*, carried a special State Department envoy, William I. Buchanan. After receiving Gómez' December 21 telegram, Secretary Root instructed Buchanan, a veteran diplomat in Latin American affairs, to meet the *North Carolina* at Hampton Roads, Virginia, and to proceed to Caracas. Buchanan was told to obtain Gómez' commitment to arbitrate the pending claims in exchange for the reestablishment of diplomatic relations. Gómez wanted the support of the United States and, after a series of tedious negotiations, settled with Buchanan. He returned the New York and Bermúdez Company properties, although he fined the asphalt company $60,000 for complicity in the Matos conspiracy, and he agreed to submit other pending claims to arbitration. A satisfied United States restored relations on March 9, 1909.[25]

By late 1908, the United States was prepared to deal with Gómez. As early as January 1907, the State Department had learned confidentially that Gómez promised, if he became president, to settle foreign claims and "to do everything possible to attract foreign capital to Venezuela." In the spring of 1908, anti-Castroites reportedly had approached the department to inquire about its attitude toward a possible revolution. Secretary Root, citing international law, refused to negotiate with the insurgents. Nonetheless, the department, which had received reports of palace intrigues in the days after Castro's departure, expected a revolution and prepared to aid Gómez. On December 17, two days before the *golpe*, President Roosevelt, noting that "things are evidently on the verge of a complete upsetting in Venezuela," alerted Root. Indeed, the warships made for La Guaira immediately after Gómez requested them. And Root gave comprehensive instructions to Buchanan on how to handle the various claims on the day Gómez offered to settle. It was, as the *Nation* gibed, "a nicely timed revolution." Roosevelt and the State Department had learned from the Panama embarrassment of 1903 to wait until the revolution began before despatching the navy.[26]

After completing his negotiations with Gómez in late February, Buchanan left with the U.S. warships from La Guaira. Both Roosevelt and Gómez had achieved their objectives. After several years of frustration, Roosevelt had purged the hemisphere of the Castro nuisance. He now had a Venezuelan ruler who promised to pay his debts and thereby keep his country from becoming a source of international disputes. He had attained that goal without having to use military force against Venezuela. His diplomatic and military support of the December 19 *golpe* was just the sort of visible approval Gómez believed that he needed to discourage counterrevolutionaries. In Gómez' words, the U.S. warships were a "convenient presence."[27]

One Venezuelan refused, however, to accept this turn of events. Ex-President Castro planned to regain his office. The incision from his operation had hardly healed when, in late March 1909, he boarded a steamship at Bordeaux, France, and announced to newsmen that as "a man of destiny" he would return to Venezuela and overthrow Gómez. Like another famous dictator, Castro fancied himself returning from Elba.[28]

The reports of Castro's intentions alarmed the new administration of William Howard Taft. The ex-dictator might have enough support to topple Gómez. Two days before the State Department learned of Castro's embarkation, the new U.S. minister in Caracas, William W. Russell, had informed Washington that Gómez had not yet rallied the population against Castro and that many of Castro's followers were in Caracas or headed there. The Taft government quickly prepared to preserve Roosevelt's diplomatic triumph of the previous December. On March 29, Secretary of State Philander C.

Knox instructed Secretary of the Navy George von Meyer to send a warship back to La Guaira to guard Gómez. The next day Knox ordered U.S. representatives to inform France, Great Britain, and the Netherlands that the Unites States would protect its interests "and that we feel in the present circumstances the position of Mr. Castro nearly approaches that of an outlaw and that those Governments which for financial, geographical or other reasons are interested in the maintenance of a responsible government in Venezuela should take all legitimate measures to prevent their ships or territory from being used in a way which would hamper the Government of Venezuela in dealing with the menace of Castro's return."[29] In effect, Knox promised to keep Castro from Venezuela and requested the Europeans to exclude him from their Caribbean possessions.

When Castro reached the Caribbean in early April 1909, on the steamer *Guadeloupe*, he found a concert of powers opposed to him. The United States blocked his return to Venezuela, Great Britain refused him entry to Trinidad, and France expelled him from Martinique. Castro had little opportunity to land anywhere else for the United States Navy shadowed his movements. Finally, on April 10, the frustrated Castro boarded another ship and returned to Europe. He remained in exile until his death in 1924.[30]

After thwarting Castro's return to Venezuela, the Taft administration relaxed its attention toward Venezuela. Gómez represented everything Castro had not been. The general proved to be an efficient tyrant who, while consolidating his rule, reorganized Venezuela's finances, regularly made payments on the international debt, and praised the United States. In turn, President Taft commended Venezuela for paying its debts in his annual message to Congress, and the State Department approved the sale of a navy gunboat to Venezuela for protection against future Castro incursions.[31] Relations between the two countries became routinely cordial. The highpoint came in April 1912, when Secretary Knox visited Venezuela as part of his tour of Latin America to celebrate the near completion of the Panama Canal.[32] In 1906, the last high U.S. official to tour South America, Secretary of State Root, had not honored Venezuela because of President Roosevelt's displeasure with Castro. Knox's visit signified the change that Gómez had brought about in Venezuelan-U.S. relations.

Commercial relations were similarly uneventful. During Taft's presidency, the United States became Venezuela's chief trading partner. But the trade was relatively small; between 1910 and 1914, the annual value of U.S. exports to Latin America averaged $302 million, but they averaged only $4.5 million to Venezuela (see Appendix, Table A). The National Association of Manufacturers' warehouse display had not generated any significant new trade, and it was closed in 1901. Venezuela still bought more than two-thirds of its imports from other countries, particularly Great Britain and Germany. Moreover, U.S. businessmen made no new substantial investments in Venezuela. The Orinoco Company actually withdrew, deciding to forego its claims to lands on the Orinoco River's west bank in exchange for money from Gómez. The company's endless litigation with Castro's courts had nearly bankrupted it. U.S. direct investments in the South American nation dwindled to about $3 million (see Appendix, Table B).[33] President Cleveland's vision of prosperous U.S. markets in northern South America had vanished.

While Venezuelan-U.S. commercial relations remained sluggish, the Gómez regime had set Venezuela upon a momentous course. Along with satisfying creditors, President Gómez welcomed foreign investors to his nation. In late 1909, he granted John Allen Tregelles, the representative of a British company, petroleum exploration rights in twelve of Venezuela's twenty states, an enormous lease of 27 million hectares. Gómez needed to meet administrative expenses and to refund Castro's debts; he hoped that the foreign development of the nation's resources would immediately bring him revenues through exploration taxes and royalties. But Tregelles' enterprise, the newly organized Venezuelan Oilfield Exploration Company, disappointed Gómez. The company had limited financial resources and, after a cursory survey of eastern Venezuela, let the concession lapse in late 1911.[34]

What Gómez wanted was a foreign corporation capable of vigorously carrying out the exploration and production of Venezuela's resources. Three weeks after the demise of the Tregelles concession, Gómez granted a prospecting license for roughly the same territory to Rafael Max Valladares, the attorney for General Asphalt, the parent company of the New York and Burmúdez Company. With Gómez' blessing, Valladares immediately transferred the concession to the Caribbean Petroleum Company, a new subsidiary of General Asphalt. In 1912, Caribbean Petroleum spent several million dollars exploring the grant, and its chief geologist, Ralph Arnold, reported favorable prospects for commercial petroleum production in the vicinity of Lake Maracaibo. But General Asphalt lacked funds to begin exploitation, and, to prevent the lapse of the concession, the firm transferred the concession in January 1913, in a complex stock exchange deal, to a British-Dutch combine, the Royal Dutch–Shell Oil Company.[35] With the rich and powerful Shell operating in Venezuela, Gómez finally had a corporation capable of transforming Venezuela's economy and making him a wealthy man.

Royal Dutch–Shell's entrance into Venezuela in 1913 marked the beginning of the nation's modern economic history. The company vigorously expanded its operations in Venezuela. After purchasing the original Tregelles concession, it sought other promising oilfields and gradually gained control over small British oil companies in Venezuela. Its subsidiaries extensively explored Venezuela and selected parcels for drilling. In February 1914, one subsidiary, Caribbean Petroleum, drilled Venezuela's first commercial well at Mene Grande, just east of Lake Maracaibo. By 1917, Caribbean's production reached 121,000 barrels. That year it shipped the first petroleum from Venezuela.[36]

The enthusiastic reports of geologists convinced Shell to invest in Venezuela. But Shell's quickened interest in Venezuelan petroleum also derived from the stability of Gómez' rule. For example, the concessions of Colon Development Company and Venezuelan Oil Concessions, two Shell subsidiaries, had existed since 1907. Castro had parceled out subsoil exploration rights to his cronies, expecting that they, in turn, would sell them to foreigners. But the vagaries of Castro's regime discouraged foreign investors, so the concessions remained unsold until 1915.[37] In addition, Gómez' Venezuela became alluring as Shell's other Latin American area of investment, Mexico, entered an era of revolution and uncertainty after the overthrow of dictator Porfirio Díaz in 1911. During his more than three decades of absolute rule, Díaz alienated virtually the entire national domain to foreign investors. Gómez, in effect, re-created the *porfiriato* in Venezuela by establishing an order that kept foreign investors safe from revolutionary activity and the threat of nationalization.

In exploring Venezuela's oilfields, the British not only expressed their confidence in Gómez and Venezuela's potential but also demonstrated their foresight. Before the outbreak of World War I, British officials had decided that petroleum was a strategically vital raw material. Indeed, the Admiralty had already converted the warships of the Royal Navy, the centurions of empire, from coal-burners to oil-burners. Since Britain had scanty petroleum reserves within its empire, London encouraged its nationals to explore throughout the world, and it even entered the oil business by purchasing controlling shares in the Anglo-Persian Oil Company. By the end of the war,

British oil companies had cornered more than half of the world's known reserves in such far-flung areas as the Netherlands East Indies, Burma, the Middle East, Rumania, and Venezuela.[38]

The United States, on the other hand, had not developed a coherent oil policy before the war. In 1915–1916, geologists of Standard Oil of New Jersey toured Venezuela, but Standard executives decided that Shell had obtained all available concessions. In any case, U.S. oilfields were rich, producing about 70 percent of the world's oil supply. Within its borders and in nearby Mexico, the United States had seemingly abundant petroleum reserves. Thus, U.S. officials did not urge the oil industry to expand its holdings, and between 1913 and 1918, they expressed no deep concern over Shell's entry into Venezuela.[39]

While British companies were striking oil in the Caribbean nation, relations between the United States and Venezuela generally were remaining commonplace. During the first years of Woodrow Wilson's presidency, only two unusual incidents occurred. The first related to Wilson's hemispheric policies. After taking office, Wilson pledged a new Latin American policy for the United States. In particular, he promised to end Taft and Knox's "dollar diplomacy." Their unabashed promotion of U.S. capital had led to armed intervention in Nicaragua and financial control over Haiti.[40] Making a memorable address in Mobile, Alabama, in October 1913, Wilson outlined his ideas. He rebuked the Taft administration, noting that "it is a very perilous thing to determine the foreign policy of a nation in the terms of material interest." He predicted that Latin America would soon escape foreign economic domination and enter "the main field of modern enterprise and action." Furthermore, he promised that the United States would never seek an "additional foot of territory" and would base its foreign policy on "morality and not expediency" and recognize only those Latin American governments that truly represented their people.[41]

In his Mobile speech, Wilson intended to explain not only his Latin American policy in general but also his Mexican policy in particular. Wilson had refused to recognize Mexican dictator Victoriano Huerta. Huerta, a spokesman for Mexico's propertied groups, had deposed the liberal reformer, Francisco Madero, in February 1913. Wilson judged Huerta a usurper opposed to the interests of the Mexican masses. A month after his Mobile speech, Wilson openly insisted that Huerta resign. And, in February 1914, he lifted the existing arms embargo to allow war material to reach Huerta's leading opponents, Venustiano Carranza and Francisco Villa.[42]

Wilson's aggressive policy toward Huerta worried Gómez. After grabbing power in December 1908, the laconic general from Táchira arranged to have himself constitutionally elected. Once cloaked with legitimacy, he exhibited the cunning, brutality, and rapacity that would characterize his rule of nearly three decades. In addition to appeasing foreign powers, Gómez tied influential Venezuelans to his regime by parceling out government posts and patronage to regional *caudillos* and marrying his sisters and innumerable children into wealthy Venezuelan families. He also masterfully manipulated traditional regional rivalries, particularly the outer states' antipathy for Caracas, to his advantage. These political maneuvers were backed by force. He secured his hold on the army, the fountainhead of power in Venezuela, by appointing only clansmen and *andino* henchmen to military posts. An elaborate network of spies and a dreaded penal system supplemented Gómez' power and terrorized individual critics. And loyalty to the Gómez system was assured by permitting officials to engage in graft, corruption, and even extortion. As for Gómez, who had been a small-scale *hacendado* in his native state, raising cattle and growing coffee, the presidency offered the opportunity to turn the nation into his private *hacienda,* as he expropriated land and cattle and impressed peasants.[43] The problem for Gómez, however, was that the constitution prohibited a president from succeeding himself and mandated presidential elections for April 1914. He needed a scheme to circumvent the constitution while still preserving the fiction of legal authority and without provoking the United States.

On August 4, 1913, Gómez imposed martial law and suspended constitutional liberties because of alleged revolutionary activity throughout the nation, but particularly in the western state of Falcón. While taking care to act within the constitution and assuring the United States he would respect Venezuelan law, the president plotted against his people. A brother-in-law of Cipriano Castro had unsuccessfully tried to convince the military commander of Falcón to lead the state in rebellion against Gómez. This and other minor disturbances served as convenient pretexts for utilizing an article of the constitution that allowed, during national emergencies, for a provisional president. Gómez maintained martial law, canceled elections, and installed a puppet, Victoriano Márquez Bustillos, as provisional president on April 19, 1914. A year later, Congress, firmly under Gómez' control, proclaimed a new constitution and "unanimously" offered the presidency to the commander-in-chief of the armed forces, Gómez. The Venezuelan had solved the problem of *continuismo,* the issue that had toppled Porfirio Díaz.[44]

Washington was not fooled by Gómez' chicanery. When martial law was declared, the State Department ordered the warship *Des Moines* to the coast of Falcón to investigate and discovered the ruse. Moreover, as early as January 1914, department officers had decided that "although President Gómez has taken care to be strictly legal in everything that he has done . . . great doubt exists as to his good faith and that very probably he is scheming to retain the Presidency himself or to transfer it to someone of his friends upon whom he can rely."[45] Yet, the Wilson administration declined to apply the Mexican nonrecognition policy to Venezuela. On April 13, 1914, for example, it received from a minority of Venezuelan congressmen, including General Guiseppi Monagas, the grandson of a former Venezuelan president, a petition requesting U.S. intercession to guarantee free elections. Special Counselor Robert Lansing not only promptly rejected the petition but also emphasized to the U.S. minister in Caracas, Preston McGoodwin, "that you must be very careful to avoid any appearance of taking any part whatever in the political affairs of the country to which you are accredited."[46]

For reasons of diplomacy, the United States acquiesced in the Gómez dictatorship. Commercially and strategically, Venezuela was less important to the United States than was Mexico, and officials thought that diplomatic intervention would not be worth the problems and turmoil it would engender. Wilson's Mexican policy had already provoked nationalist backlash in Mexico and strident criticism in the United States and Latin America. Furthermore, field officers reminded the Department of State that Gómez was "inclined to be pro-American" and that he had a firm grip on the country.[47] Finally, officials concluded that they would be unable to find an acceptable alternative to Gómez and that chaos might ensue if he fell from power. As one department official had noted: "It seems to me we must be very careful about handling Gómez. We are morally, if not more so, responsible for putting him in there, and while we must certainly look with disfavor on any perpetuation of unconstitutional methods in the sister Republics, it is an open question to my mind as to whether Gómez with all his faults may not be better than some weak creature. . . . The only thing that can do any good in Venezuela is a moderate soft-shell despotism."[48] With its decision to reassert the traditional U.S. foreign policy of recognizing de facto governments, the Wilson administration avoided a potentially explosive issue.

The second Venezuelan-U.S. incident came during World War I. Venezuela was neutral throughout the war. Venezuela's neutrality was not unusual, for, although seven Central American nations declared war against the Central Powers after April 1917, only one South American nation, Brazil, allied with the United States. What was unusual was that Venezuela sympathized with Germany. Unlike Venezuela, the other South American nations either kept silent or spoke favorably of the Allied cause.[49]

General Gómez admired Germany. His motto for Venezuela was "union, peace, work"; he believed that Germany's rapid industrialization in the late nineteenth century demonstrated what could be accomplished under disciplined rule. Moreover, as a military man, Gómez respected Germany's military efficiency and prowess and approved of the position that its army achieved in German political life. On a less philosophical level, Gómez understood Germany's economic role in his country. Germany was a major trading partner, and its capitalists controlled many of Venezuela's banks and international trading firms. Finally, the Venezuelan had a personal stake in a German victory. Before the war, he had invested $2 million in Germany.[50]

Although admiring Germany and predicting that it would be victorious, Gómez maintained Venezuela's neutrality during the war. In mid-1917, for example, he denied in writing to the State Department rumors that he was contemplating selling Margarita Island, a small stretch of land off Venezuela's central coast, to Germany and pledged to prevent any hostile action against the United States from emanating from his nation. His sentiments were displayed, however, by wearing a Prussian-style uniform, suppressing pro-Allied newspapers, and jailing pro-Allied journalists. These domestic policies incensed the Wilson administration and reminded it of Gómez' manipulation of the constitution and that he ruled through "a policy of terrorism."[51] As the war raged in Europe, a Venezuelan-U.S. confrontation loomed.

Beginning in late 1917, the State Department considered courses of action against Gómez. In November, William W. Russell, the former minister to Venezuela, informed Secretary of State Lansing of his conversation with a Venezuelan exile, José María Ortega Martínez. Ortega Martínez strongly advised against any hostile action, for "any pressure brought by the United States either separately or in concert with other nations, against Gómez, would immediately be used to arouse the patriotism of Venezuelans against foreign interference and *would react* seriously against us." In Ortega Martínez' opinion, the policy to pursue would be to arm covertly anti-Gómez exiles. Department officer Glenn Stewart followed Russell's memorandum with a lengthy analysis of the Venezuelan problem. Stewart suggested three possible policies toward Gómez: nonrecognition, a commercial embargo, or strong diplomatic pressure. Stewart rejected all three alternatives. A nonrecognition policy would fail; Gómez was financially and economically independent. An embargo would be similarly unsuccessful, for Venezuela did not have an export tax, and, in any case, the U.S. war effort could use Venezuelan coffee, sugar, cacao, and hides. Finally, Stewart agreed with Ortega Martínez that strong diplomatic pressure would only redound to Gómez' benefit. Stewart concluded "There is only one way to guarantee good Government and Christian civilized conditions. Eliminate Gómez root and branch."[52]

Secretary Lansing heartily approved of Stewart's recommendation. He sent the memorandum to President Wilson with the notation that Gómez was a "real danger to the United States and our War effort." On February 16, 1918, Wilson replied to Lansing: "This scoundrel ought to be put out. Can you think of any way in which we can do it that would not upset the peace of Latin America more than letting him alone will?"[53] Wilson did not follow up his outburst. The administration's only hostile action was an embargo of newsprint to Venezuela.[54] As with the recognition issue of 1913–1914, U.S. officials probably decided that a solution to the Gómez problem would be more troublesome than the problem itself. To "put out" Gómez would mean direct intervention, perhaps requiring the landing of marines or the arming of exiles combined with naval support for them.[55] In early 1918, the Wilson administration did not need another war or domestic criticism of an interventionist policy.

As U.S. troops began to break the military stalemate in Europe in the summer of 1918, the Department of State reconsidered Venezuelan relations. On September 4 the Latin American Division told Secretary Lansing that it had been puzzled by Gómez' hostility but that, after careful study, it concluded that German commercial power in Venezuela and German military prowess influenced his attitudes. The division emphasized that General Gómez was not pro-German, but pro-Gómez. If the United States countered German influence by demonstrating its strength, the Venezuelan would support U.S. policies. Accordingly, the division recommended "that accurate information should be given to General Gómez regarding the military power of this country."[56] Germany's military collapse two months later, however, made a propagandistic campaign unnecessary. And, as the division predicted, Gómez promptly abandoned his love for Germany and aligned his nation with the triumphant powers, particularly the United States.[57]

Though causing consternation with the U.S. government, neither Gómez' subversion of the Venezuelan constitution nor his pro-

German sentiments precipitated a diplomatic clash. In both cases, the Wilson administration considered coercing Gómez into embracing its hemispheric and world policies. But it eventually recognized that Gómez' behavior posed no threat to its international interests. Without the postwar energy crisis, the Wilson government and succeeding administrations probably would have continued to treat lightly diplomatic relations with Venezuela. As during the Roosevelt and Taft presidencies, their major concern would have been to prevent Venezuela from becoming, as one official put it, "an arena of friction . . . in connection with the Monroe Doctrine and the hegemony of the United States in this hemisphere."[58] The newly perceived need for petroleum, however, encouraged the United States to give priority to relations with Venezuela.

2. Open Door Diplomacy, 1919–1929

Prior to 1919, U.S. businessmen had little interest in Venezuela, an impoverished agricultural nation. Following the end of World War I, however, government officials grimly concluded that the United States had nearly exhausted its domestic petroleum reserves. Oilmen, with the aid of the Department of State, responded to the postwar energy crisis by supplanting British oil companies in Venezuela and establishing U.S. economic hegemony there. And, as oil money flowed into Venezuela, the United States actively supported the vicious and venal regime of Juan Vicente Gómez.

A crucial lesson of World War I was that petroleum was indispensable for an industrial nation's defense. Beyond being the primary source of energy for modern navies, petroleum fueled the new war machines: the airplane, submarine, supply truck, and tank. Throughout the war, both the Allies and the Central Powers often became dangerously deficient in that vital raw material. As the French leader Georges Clemenceau anxiously told President Woodrow Wilson in December 1917, "petrol . . . is as necessary as blood in the battles of to-morrow."[1] In addition to its strategic significance, petroleum's commercial possibilities, as with the gasoline-powered automobile, were becoming obvious.

Fortunately for the Allies, oilfields in the United States produced record quantities, supplied 80 percent of the war requirement, and averted a possible catastrophe. But the U.S. oil industry's prodigious wartime effort led to postwar fears. Geologists believed that the United States had nearly exhausted its domestic petroleum reserves. In 1919, the director of the United States Bureau of Mines, Van H. Manning, warned that U.S. oil production would peak in two to five years and then decline. The country's wells were producing over 350 million barrels a year, a full one-twentieth of the nation's known reserves. As George Otis Smith, director of the United States

Geological Survey, gloomily observed in January 1920, "the position of the United States in regard to oil can best be characterized as precarious."[2]

U.S. officials became further alarmed when they realized that, while they had complacently banked on reserves in the United States and Mexico, British oil companies had aggressively expanded their holdings over the globe. In 1919–1920, U.S. alarm turned to anger when the Department of State learned that the British were excluding U.S. oil companies from the most promising oilfields, the mandates of Mesopotamia and Palestine. After 1920, the department labored assiduously to undermine the British hold on the Middle East and to open the area to U.S. capital. In July 1928, the department's persistence and pressure opened the Turkish Petroleum Company to a consortium of U.S. oil companies. The successful drive for a share of the oil rights is, of course, among the most widely celebrated and studied events in diplomatic history, for it led to U.S. economic expansion in the Middle East.[3] But in the wake of the depletion of domestic petroleum reserves, the U.S. government cast a longing eye not only on distant petroleum prospects but also on those much closer to home. In the postwar years, it attempted to pry open the door of economic opportunity in Venezuela as well as in Mesopotamia.

Responding to fears of an oil shortage, the forceful British oil policy, and the request by U.S. oil companies for diplomatic assistance, the Wilson administration began to formulate an energy policy in August 1919. In a letter to all embassy and consular officials, the State Department asked for information on oil concessions, noting that "the vital importance of securing adequate supplies of mineral oil for both present and future needs of the United States had been forcibly brought to the attention of the Department." Beyond collecting data, the department also instructed its representatives "to lend all legitimate aid to reliable and responsible citizens who are seeking mineral oil concessions." The department added the caveat, however, that before rendering diplomatic assistance the representatives distinguish between bona fide U.S. companies and those merely incorporated under U.S. laws but dominated by foreign capital.[4]

U.S. representatives in Venezuela heeded the department's urgent request. Minister Preston McGoodwin sent reports to Washington in late 1919 and 1920 on the activities of British oilmen. Consul Dudley Dwyre, "knowing that the Department realizes the vital importance of securing adequate supplies of oil for the future needs of the U.S.," advised his superiors that Archie Davis, a U.S. citizen and resident manager of the Colombian Petroleum Company, would be in Washington and could apprise them of Venezuela's possibilities. In April 1920, "in the interests of American capital," McGoodwin convinced President Juan V. Gómez to block the plan of Vicente Lecuna, the president of the Bank of Venezuela, to restrict foreign participation in petroleum exploitation, and in June he arranged a personal interview with Gómez for Sinclair Oil Company executives. For such efforts, McGoodwin won the department's praise but did not challenge British monopoly.[5] Royal Dutch–Shell still controlled all the significant petroleum concessions, and British investments in other Venezuelan enterprises, such as public utilities and railroads, far outranked those of other nations. Moreover, Gómez seemed to favor British capital. Throughout 1919 British agents in Venezuela smugly reminded the Foreign Office that Gómez both feared his northern neighbor's economic and military power and resented its wartime pique and that the Venezuelan hoped British capital would counteract U.S. influence. As British Chargé d'Affaires Cecil Dormer noted, "he [Gómez] is the best guarantee that we have that British interest will not be interfered with."[6]

The British miscalculated both the economic power of the postwar United States and the cupidity of President Gómez. The United States went from a debtor nation to the world's greatest creditor during the war. Though the change can be overemphasized, for the United States had extensive investments in Cuba, Mexico, and Central America before 1914, U.S. corporations and New York banking houses relentlessly expanded the country's international finance and trade in the decade after the war. In particular, U.S. capitalists challenged the British for the role of chief lenders and investors in South America. By 1929, the value of U.S. direct investments in South America, for example, was 840 percent higher than in 1913.[7] Throughout the 1920s the British legations in South America transmitted one discouraging report after another about the negotiations for a new U.S. loan or the establishment of a new U.S. enterprise. As one foreign officer sadly concluded after reading the latest report, "the facts are only too true, and the despatch makes depressing reading. The U.S. have money to spare; we have not; and only can wait on time."[8]

In 1919–1920, Venezuela fairly swarmed with curious and ambitious U.S. visitors. Speculators and oil company representatives bid for concessions and offered handsome sums and bribes to Gómez. They thought they spotted an opening, for Shell seemed unable to expand production. The Venezuelan oil industry's progress had slowed during World War I. Transportation shortages and import restrictions imposed by the British made it impossible for Shell to obtain machinery and other supplies. For example, one subsidiary found petroleum east of Lake Maracaibo in 1916, but it could not find the equipment to transport it. After the war, shortages persisted, and until the end of 1919, Great Britain continued wartime restrictions on the emission of new capital.[9] In June 1919, Gómez had pointedly expressed to Henry Beaumont, the British minister, his belief that with the conclusion of the war "it would now be possible for foreign capital to come to help in the development of the country and that a period of great prosperity was approaching."[10] Shell's wartime production had been promising, but in relation to world output it was, at less than a million barrels a year, insignificant. Gómez received from Shell officials glowing estimates of Venezuela's possibilities, but no one had made either him or Venezuela prosperous.

Gómez, expressing dissatisfaction with Shell's slow pace, threatened on April 7, 1920, to nullify the concession of its subsidiary, Colon Development Company. According to the terms of the concession, a company had four years to choose parcels of land, begin work, and pay surface and exploitation taxes. All unworked parcels reverted to the government. Shell, however, considered the concession one great petroleum mine and argued that work in one area meant that the entire concession was in exploitation. Yet, Shell insisted it owed taxes only on those parcels it actually worked. The tax difference amounted to paying either Bs. 1,200 (bolívares) on the lots selected or Bs. 3.8 million on the entire concession.[11] During the war, Gómez quietly accepted Shell's logic, but when U.S. money became available, he no longer was as patient.

Both the State Department and the U.S. oil industry confidently awaited the outcome of the Venezuelan suit against Shell. While the British were assuring themselves that Gómez feared the United States, the Venezuelan tyrant informed Minister McGoodwin in November 1919 "that he would welcome American capital and that he would offer 'absolutely every assurance of protection to American lives and property' even in the remote parts of the country." Moreover, the oil companies had encouraged Gómez with the promise that they would replace Shell and immediately begin exploration and production.[12] The State Department also helped by avoiding a diplomatic pitfall. Soon after Gómez' announcement, the department received an urgent message from the Carib syndicate. Carib, a U.S. concern, had a 25 percent interest in the Colon company; it implored the department to intercede in the Venezuelan suit. The department remembered, however, its August 11, 1919, circular, checked into the full ownership of Colon, and refused to support an essentially British oil company.[13] After having been outmaneuvered in the world's oilfields, the United States in the mid-1920s seemed on the verge of a diplomatic triumph. It was a point not lost on the British. After hearing of the lawsuit, Foreign Officer Ernest Weakley noted to his colleagues that "they are at present irritated by our attitude in refusing them access during the period of Occupation, to the oilfields of Palestine and Mesopotamia, and this may presumably be a reason for Standard Oil intrigues in South American States against British companies who happen to have obtained concessions in those countries."[14]

Excitement mounted in the U.S. oil industry when on June 11, 1920, Venezuela's Supreme Court annulled Colon's concession, except for the parcels which it had declared in exploitation. The ruling implied that almost all of Shell's holdings would be jeopardized, for none of its subsidiaries had done extensive drilling. In fact, the government had already initiated a suit against Venezuelan Oil Concessions. Representatives of six U.S. oil companies began negotiations for the annulled concessions.[15] Gómez had opened Venezuela's door to the north.

But to the U.S. companies' chagrin, Gómez announced in February 1921 that he had canceled the Supreme Court's decision and that Colon Development would have five years to select preliminary tracts for exploration and another five years to make a final selection and commence drilling. In return, Colon had agreed to pay a small exploitation tax on all the tracts it selected. Once again, British diplomats had circumvented the United States. In the summer of 1920, Chargé d'Affaires Dormer convinced the Foreign Office that the Venezuelans had legitimate grievances, that Shell's interpretation of its concession rights was unreasonable, and that this was one occasion when the oil executives could not bribe their way back into Gómez' favor. Dormer also wrote strong notes to the Venezuelan Foreign Ministry warning that Britain would not accept an adverse decision and that British businessmen would boycott Venezuela. In his efforts to reach a settlement, Dormer was ably assisted by the new resident manager of the Shell interests, William T. S. Doyle, who ironically had been the chief of the State Department's Division of Latin American Affairs during the Taft administration. By November, Dormer and Doyle had negotiated the basic outlines of a settlement. They had impressed the government with the seriousness of British intentions, while assuring it that Shell truly intended to produce Venezuela's petroleum. They also had prevented "hopeful Spectators" from taking advantage of an adverse decision.[16]

While the British adeptly formulated new policies, U.S. officials fumbled their assignments. After learning of the Supreme Court's June 1920 decision, the State Department reconsidered its refusal to support the Carib syndicate's claim. Noting that the "decision of the Supreme Court would apparently destroy the Carib Syndicate," the department instructed McGoodwin on June 24 to inform the Venezuelan government that the United States "would be pleased" if it revalidated Carib's share of the concession. McGoodwin, however, misinterpreted his instructions and told Gómez that the United States desired the revalidation of the entire Colon Development concession. When Gómez announced in February the reversal of the court's decision, McGoodwin proudly reported that "there can be no doubt that the government of Venezuela was actuated in reaching a settlement by its desire to protect the one-fourth American interest in the company."[17] McGoodwin's message shocked the department. As one member of the Latin American Division lamented, "it is a particularly unfortunate piece of business that this tremendous British concession should come to life again through the efforts of our minister, when it has been the particular aim of the Department to see this concession cancelled."[18] McGoodwin, however, had overemphasized his influence. After the department replaced McGoodwin, a political appointee of the Wilson administration, in late 1921, Chargé d'Affaires John Campbell White assured the department that British pressure and Gómez' desire to avoid international complications were the chief explanations for the revalidation of the Colon concession.[19]

The diplomatic triumph in February 1921 marked the high point of British economic influence in Venezuela. A thaw in Anglo-American relations, the persistence and pervasiveness of U.S. capital, and the skillful oil diplomacy of the Harding and Coolidge administrations all helped to loosen the British hold on Venezuelan oil. After a series of discussions and notes, the British removed in 1922 the chief cause of Anglo-American friction by agreeing, in principle, to open Middle Eastern oilfields to a group of U.S. oil companies. The British apparently concluded that U.S. capital, engineering skill, and diplomatic influence might after all be useful in the Middle East.[20] The Washington Naval Conference of 1921–1922 similarly eased tensions by taming the naval rivalry beween the two countries. The new cordiality was also evident in Venezuela. In April 1922, Standard Oil of New Jersey bought one-third of the holdings of a company, British Controlled Oilfields. British Controlled lacked the capital to undertake oil production and turned to the U.S. corporate giant. Two years before, the sale would have been impossible, but now, as one avid promoter of British capital, Ambassador to the United States Sir Auckland Geddes, observed, "it would seem that the conclusion of such an arrangement . . . would be beneficial to both parties and would promote cooperation between British and American oil interests."[21]

Besides purchasing British properties, U.S. oil companies bought new concessions in the 1920s. Standard Oil of New Jersey, Standard Oil of Indiana, Gulf Oil, New England Oil, Sun Oil, Sinclair Oil, and Texas Oil obtained leases. Even if they had desired to do so, the British no longer had the power or influence to exclude U.S. companies. As always, President Gómez sought to profit from the leasing of the national domain. The price paid allegedly included a "succulent commission" for Gómez and his friends.[22] Shell, already satiated with its prewar concessions, did not need any more land. And no other British oil company in Venezuela could outbid or outbribe the U.S. companies. Indeed, the British resigned themselves to the U.S. commercial intrusion. Department of Trade Officer Robert Leslie Craigie summarized official thinking: "In view of the power of the dollar, British oil interests in Venezuela will do well to compromise with their American confréres and to admit a certain American participation in their undertakings."[23]

The State Department, responding to pleas by oilmen to "bring to bear, in full measure, diplomatic and economic pressure" to secure "oil resources under the control of American capital and management," astutely assisted the oil industry's expansion. As during the Wilson administration, it arranged for oil executives to meet influential Venezuelans, including President Gómez. In addition, the legation at Caracas protested to the Foreign Ministry whenever the oil industry objected to a proposed tax or regulation.[24] The department also cooperated with British oil interests and twice joined in successfully supporting the Britons in lengthy legal disputes with Venezuelan ministers over the terms of exploration concessions, because U.S. citizens owned minority interests in the British companies. Overt support protected U.S. investments and contributed to Anglo-American goodwill; it recognized, as Arthur N. Young, a department economic advisor, put it, that the British now "played the game" in the world's oilfields.[25] Simultaneously, the department covertly opposed the entrance into Venezuela of the one oil company, besides the Royal Dutch–Shell, that had the capital, technology, and diplomatic support to compete with the U.S. companies, the Anglo-Persian Oil Company. The British government controlled 51 percent of Anglo-Persian's stock, and Venezuelan law traditionally forbade foreign governments from acquiring concessions.

Moreover, State Department officials reasoned that a government-controlled corporation in Venezuela would violate the Monroe Doctrine. After learning that Anglo-Persian representatives visited Venezuela in September 1925, the department authorized Minister Willis C. Cook, if the Venezuelans asked, to identify Anglo-Persian's stockholders and remind the Venezuelans of their laws.[26]

In pursuing these policies, the United States attempted to attain strategic objectives. Its citizens initially became interested in Venezuelan petroleum, because they believed the United States had depleted its domestic reserves. The Wilson administration and the oil industry generally ignored British activity in Venezuela prior to 1918. After the war, however, the Wilson, Harding, and Coolidge administrations made it a matter of national security to obtain new sources of petroleum. By 1924, U.S. fears of an oil shortage had begun to subside. Geologists discovered vast new fields in California, Louisiana, Oklahoma, and Texas. The oil industry more than doubled its 1918 output in 1923 and expanded production thereafter. Oilmen worried about surplus rather than scarcity. The State Department, accordingly, relaxed its promotion of U.S. oil companies in Venezuela.[27] To be sure, some government officials reasoned that, for future national security, the United States should expand its overseas holdings. The Federal Oil Conservation Board reported in September 1926 that the acquisition of new fields in South America "is of first importance" and would prevent any future "exploitation of our consumers by foreign controlled sources."[28] Secretary of Commerce Herbert Hoover similarly argued that, in order to avoid "foreign manipulators," the United States must have plentiful sources of essential raw materials "under its own flags." Hoover worried that raw-material producers might set prices arbitrarily and thereby blackmail the United States, as he believed Brazil did with its coffee valorization schemes. Hoover continually prodded the State Department to take care of the country's future needs.[29] But the State Department, which resented Hoover's meddling, retained full control over petroleum policy, and, in view of the new domestic petroleum discoveries, it practiced a less urgent oil diplomacy.

The department did not, of course, forget about U.S. oil interests in Venezuela. In June 1923, associates of Gómez announced the incorporation of the Compañía Venezolana de Petróleo. As oilmen quickly surmised, the concern was a front for Gómez and his friends. It held all the unleased national reserves and would readily sell them to foreign oil companies. Such a "setup," as British Minister Andrew Bennett pointed out, would "enable the president and his satellites to benefit directly and decorously by the sale of concessions, and the opportunity is not missed in the Presidential message to invite buyers to present themselves." But the oil companies refused to take advantage of Gómez' latest chicanery. They feared that after the sixty-six–year–old Gómez died, a new regime might cancel the leases. In any case, with the new discoveries in the United States, the oil industry had a surfeit of possibilities.[30]

Ignored by the U.S. oil industry, Gómez took the offensive. In February 1924, Compañía Venezolana announced that it had granted to a German group, the Hugo Stinnes interests, a forty-day option to purchase 200,000 hectares of national reserves. The announcement was a bluff. Gómez hoped to deceive the oil companies into believing that their reluctance to deal with Compañía Venezolana would introduce a powerful new competitor. Gómez succeeded in his ruse, for by late April, Standard Oil of New Jersey began to purchase leases. Standard decided that, even if the leases were of questionable legality, they could plead the "doctrine of innocent purchaser," if a new administration attempted to nullify them. Once Standard began to buy from Compañía Venezolana, the Stinnes group disappeared.[31]

The State Department also participated in the Compañía Venezolana–Stinnes imbroglio. It reacted to news of the proposed option to lease by telling Chargé Frederick Chabot that, "if you believe action imminent either (a) violating existing rights of United States companies or (b) calculated to create foreign monopoly of development of national petroleum reserves, inform Venezuelan officials that this Government would view with concern any action suggestive of confiscation or which exclude United States interests from opportunity to compete on basis of equality with other foreign interests for acquisition of concessions in future development of areas constituted national reserves."[32] As it had in Mesopotamia, the State Department insisted that the door of economic opportunity must be kept open to U.S. investors and enterprises. Chabot relayed his instructions to the Foreign Ministry in late March 1924 and received a reply from Foreign Minister Itriago Chacín that, while not conceding the point on equality of opportunity, assured the United States that Venezuela would respect U.S. property.[33] Standard's decision to negotiate with Compañía Venezolana removed, of course, the department's fears that U.S. oilmen would be unable to obtain new concessions.

In insisting upon the doctrine of the "Open Door," the principle that nations should have equal access to trade, investment, and raw materials, Secretary of State Charles Evans Hughes reiterated the policies of Woodrow Wilson's administration. In 1919–1920, the department had decided that the way to gain access to the world's oilfields would be to insist upon reciprocity and the Open Door; it had noted that the British replied to complaints about Mesopotamia with allegations that the United States discriminated against other nations in the Philippine Islands and Central America. In February 1921, petroleum expert Arthur C. Millspaugh had summarized the department's position in a lengthy memorandum for the incoming administration. Millspaugh concluded that the United States should have one consistent oil policy for the entire world and that it should favor equal treatment and opportunity for all and oppose monopolistic practices.[34] But the department failed to adhere to its policy when "the particular aim of the Department" was to have Venezuela annul the Colon Development Company's concession.

The department continued to compromise its Open Door policies throughout the 1920s. After relaying the department's March 1924 message, Chargé Chabot presented his superiors with a new problem. He feared that Standard Oil's decision to purchase concessions from Compañía Venezolana might lead to the creation of a new U.S. monopoly in Venezuela. Secretary Hughes clarified the department's position for Chabot by assuring him that the department believed in equality of opportunity and would not support an attempt by either the U.S. oil industry or an individual firm to obtain a monopoly. He noted, however, that "the first concern of the Department" would be to protect the "established and actual rights of the American companies." The department would not act unless a "violation of the lawful rights of an American interest has been or is about to be perpetrated," as it initially feared had happened in the Stinnes case. The department assumed that Venezuela was "acting in good faith," but to ensure that all understood the department's thinking, Hughes instructed Chabot to approach Venezuelan officials informally and assure them that the United States was confident that their government would observe the principle of equality of opportunity.[35] To his superiors' puzzlement, Chabot remained curious about their policies and worried about events in Venezuela. He had learned from Major Thomas R. Armstrong, Standard Oil's representative in Venezuela, that Standard would not mind having a monopoly in Venezuela and over 50 percent control of Compañía Venezolana. Hughes replied that majority ownership of a national reserves company did not constitute a monopoly: "the test there would be whether the concessions held by such a company is a monopoly." If that was the result, the secretary of state reiterated that, in principle, the department opposed monopolistic concessions. But, "it does not follow that the Department can restrain American companies from negotiations which are inconsistent with its views."[36]

In contrast to its declared impotence in the case of a U.S. monopoly, the department prepared eight months later to block "a monopoly of foreign interests." In January 1925, Minister Willis Cook reported that Royal Dutch–Shell had unsuccessfully attempted to purchase a monopoly share of Compañía Venezolana. The department expressed satisfaction at Shell's failure and entreated Cook to inform it immediately "should any apparently critical situation arise which might threaten to exclude the nationals of this country from equal opportunities to participate in the development of the petroleum reserves in Venezuela." In April, the new secretary of state, Frank B. Kellogg, repeated those instructions when he told Cook to protest to the Venezuelan Foreign Ministry should Shell reopen negotiations with Compañía Venezolana.[37]

Despite a public commitment to the Open Door policy, the United States pursued in the 1920s an economic policy that had the effect of establishing U.S. economic hegemony in Venezuela. After 1922 and the general improvement in Anglo-American relations, the United States accepted the initial economic predominance of Royal Dutch–Shell in Venezuela. But the State Department refused to countenance any new foreign monopolies. It objected to the designs of both the Stinnes interests and Shell in their presumed dealings with Compañía Venezolana, and it discouraged Anglo-Persian's entrance into Venezuela. Yet the department tolerated U.S. monopolies. As Standard Oil of New Jersey purchased one embezzled lease after another, the department confessed its inability to restrain its nationals or, at best, made informal representations to Venezuelan officials. The U.S. government and corporations actually practiced, as Joan Hoff Wilson has described it, a "Closed Door" policy in Venezuela.[38]

Even if the Department of State had adhered strictly to the Open Door, U.S. oilmen would have succeeded in Venezuela. The Open Door presupposed that all nations competed, fairly and evenly, in the world's marketplace. But in a postwar world of equal economic opportunity, the strongest competitors had advantages. Before it commenced drilling, an oil company had to cut roads, establish supply lines, and construct homes, hospitals, and schools. One U.S. company claimed, for example, that it had spent over $40 million in eastern Venezuela before it had ever marketed its first barrel of oil.[39] Very few nations had corporations capable of financing such undertakings. When it insisted on the Open Door in Mesopotamia or in Venezuela, the United States was simply pointing out that, as a strong and rich competitor, it had a right to join in the development of new sources of petroleum. Once U.S. corporations had obtained oil concessions in the Middle East and Venezuela, the Open Door policy was shelved.[40]

The U.S. oil industry not only successfully entered the Venezuelan market but also reaped great benefits from Venezuela's oilfields. In 1920, Venezuela still produced less than a million barrels of petroleum a year. By 1928, Venezuela exported more crude oil than any other nation and ranked second in world output, with an annual production of 106 million barrels (see Appendix, Table C). The key event that fulfilled geologists' predictions occurred on December 14, 1922. Venezuelan Oil Concessions drilled deeper in an abandoned well one and one-half miles from the eastern shore of Lake Maracaibo in the state of Bolívar. The well, Barroso Number 2, "blew in" at 1,500 feet. It flowed initially at 2,000 barrels a day but then increased rapidly until it flowed wildly at 100,000, destroyed the derrick, and blew a column of petroleum 200 feet into the air. The gusher, reputed to be the most productive in the world, obviously convinced the oil industry of Venezuela's potential.[41]

Though Barroso Number 2 flowed on a Shell concession, U.S. oil companies profited, because it indicated that a great reservoir of petroleum existed in the Maracaibo Basin. Lago Petroleum Company, soon to be a subsidiary of Standard Oil of Indiana, obtained from Gómez' friends lake-bed concessions near the Barroso well, sank wells on the edge of its lease, and became in 1924 the first U.S. company to export Venezuelan oil. Another U.S. firm, the Venezuelan Gulf Oil Company, a subsidiary of Gulf Oil of Pittsburgh, acquired a narrow marine-zone strip of land between Lago's lake-bed concession and Venezuelan Oil Concessions' land holdings and began to export in 1925.[42]

With Gulf, Shell, and Standard of Indiana, Venezuela had three of the world's five largest oil companies operating in its fields. With their extensive financial resources and ability to command the best in engineering skill, the three companies overcame the logistical problems of working in a pioneer area and nearly doubled Venezuelan production every year from 1922 to 1928. No other oil company had any commercial success in Venezuela. More than a hundred companies, mostly from the United States, prospected in Venezuela, but small companies simply could not afford the initial investment required for exploration and drilling. In the 1920s, Gulf, Shell, and Standard of Indiana produced more than 98 percent of the South American nation's petroleum.[43]

During the boom period, the U.S. oil industry became preeminent in Venezuela. With both Gulf and Standard of Indiana working the Bolívar coastal fields, U.S. companies increased their share of Venezuelan production from less than 5 percent in 1924 to 31 percent in 1925. Thereafter, the U.S. share steadily grew with Gulf and Standard of Indiana's combined total production surpassing Shell's, for the first time, in 1929. After 1929 and until 1976, the U.S. oil industry always produced more than half of Venezuela's prodigious output.[44]

The future giant of the Venezuelan oil industry, Standard Oil of New Jersey, failed to produce commercially any petroleum in the 1920s. It drilled extensively but usually ended up with dry holes. Standard obtained, however, most of Venezuela's unleased land. By agreeing to deal with Gómez' Compañía Venezolana in 1924, Standard won the dictator's favor. From 1925 to 1935, Gómez granted Standard national reserve lots, even when the New Jersey firm's bid was lower than others. Then the company's drilling luck began to change, and by the mid-1930s it surpassed Shell as Venezuela's leading producer, and by 1945 Standard Oil of New Jersey produced more petroleum than all other companies combined.[45]

During the 1920s, U.S. businessmen gained control over not only Venezuela's oil but also its trade. During World War I, normal patterns of trade were disrupted by submarine warfare and naval blockades. The beneficiaries were U.S. exporters who eagerly moved into the Venezuelan marketplace. Their share of the market grew from 39 percent in 1913 to almost 70 percent in 1917. And European businessmen found that, when they returned with their wares after the war, they were unable to supplant their U.S. competitors. Having once established themselves, U.S. businessmen effectively employed some old and new trading advantages. Geographical proximity guaranteed competitive shipping costs, and the new Panama Canal enabled West Coast entrepreneurs to join in the trade. Furthermore, with capital available in New York money markets, U.S. exporters could grant credit on new consumer items, such as automobiles and electrical appliances, whose mass production U.S. industries had pioneered. The influence of the German merchant community, the traditional source of credit in Venezuela, was therefore sharply reduced. In the 1920s, U.S. businesses provided Venezuela with over one-half of its imports, as Venezuela increased its imports fourfold over prewar levels. Sales were still, however, relatively modest; the trade in 1929, a peak year, was approximately $45 million, which represented less than 5 percent of U.S. exports to Latin America. Nevertheless, the new pattern of U.S. domination of the Venezuelan market would persist and that market would grow increasingly lucrative.[46]

The growth of U.S. investments in Venezuela overshadowed trade developments. In 1918, the United States had a small stake in the South American nation, certainly less than $5 million. By 1929, direct investments totaled over $230 million, with the rich oilfields attracting virtually all the new capital. Six other Latin American countries—Cuba, Mexico, Argentina, Brazil, Chile, and Colombia—had more U.S. investments than Venezuela. But the pattern of investment in Venezuela was unique, for U.S. money was in direct investments, while one-third of the $5.1 billion invested by U.S. citizens in the rest of Latin America was in indirect investments, such as government bonds and securities.[47]

The rapid expansion of U.S. trade and investment in Venezuela testified to the opportunities for gain in that nation. The U.S. oil industry operated under the oil law of 1922, a code which they helped write and one which they predictably praised "as favorable as that of any country." The law provided liberal terms for the companies. Their fees included a one-tenth bolívar per hectare exploration tax, a fixed two bolívar per hectare exploitation tax, and a two bolívar per hectare surface tax. Royalties on production ranged from 7.5 to 10 percent on the market price of crude oil, less transportation costs. In comparison, landowners in the United States normally received a 12.5 percent royalty on crude oil production. Oilmen argued that Venezuela was an unproven, underdeveloped area and that they would have to invest large sums of money with no guarantee of success. These risks justified comparatively low taxes and royalties. Yet, Bolivia's potential oilfileds were in similarly remote areas, and in 1922 the Bolivian government obtained a contract with Standard Oil of New Jersey gaining a royalty of 11 percent and a progressive exploitation tax. And Argentina in 1922 created Yacimientos Petrolíferos Fiscales, a state petroleum agency with supervisory functions over foreign activity and exploitation powers of its own. Gómez, perhaps knowing he would garner bribes from the leasing of the national domain, made an easy bargain with oilmen. During his rule, the government probably collected no more than 15 percent of the profits from the sale of a nonrenewable natural resource.[48]

Operating under such a generous law, the foreign oil companies made enormous profits. Venezuela's oil boom coincided with a growth in world demand for petroleum and its by-products, particularly gasoline and motor oil. In 1927, Standard of Indiana's subsidiary, Lago Petroleum, reported, for example, earnings of $8 million on a working capital of only $3.5 million. Shell's subsidiary, Venezuelan

Oil Concessions, similarly flourished with a 111.5 percent return in 1928 on an investment of one million pounds sterling. Total industry earnings during Gómez' rule may have reached $1,666 million.[49]

In the 1920s, as the interests of the U.S. oil industry and Venezuela became intertwined, the United States bolstered the regime of Juan Vicente Gómez. The Harding and Coolidge administrations sent friendly telegrams to Gómez and participated in public demonstrations of Venezuelan-U.S. friendship, such as unveilings of statues and laying of wreaths. They also dispatched U.S. heroes, like General John Pershing and aviator Charles Lindbergh, to Gómez' court. Moreover, the United States prepared to support the Venezuelan in a crisis. On June 1, 1923, Minister Cook informed the State Department that he had heard rumors of an impending revolution in Venezuela. Cook's message alarmed the department. On June 13, Secretary of State Hughes instructed Secretary of the Navy Edwin Denby to prepare the Special Service Squadron for a visit to Venezuela. When he learned of Hughes' action, Cook asked him to postpone the visit of the flotilla, for he had no evidence of a revolution; he had heard only rumors. On July 9, Cook changed his mind. He urged that the commander of the Special Service Squadron make a "courtesy visit" to Venezuela. Reliable sources had told him of impending upheavals. A visit would, in Cook's opinion, "have a quieting effect on the country and is not objectionable to the Government."[50]

The crisis passed quickly. The rumors arose after the mysterious death of Gómez' brother, Vice-President Juan Crisóstomo Gómez. The ruler's son, José Vicente, was responsible for the murder and Gómez probably created the rumors of revolution to cover his son's crime.[51] In any case, the Special Service Squadron did not immediately need to embark for Venezuelan waters. The squadron's flagship, the *Rochester*, eventually called at Venezuelan ports in mid-September to demonstrate U.S. friendship.[52]

U.S. foreign policy toward Gómez had come half-circle in the five years since the end of World War I. In mid-1923, the State Department was willing to use the navy to support Gómez. The displays of U.S. military might were no empty gestures. In the 1920s, the United States dominated the Caribbean; its troops ruled Haiti, Nicaragua, and Santo Domingo. A naval demonstration would warn revolutionaries that the United States opposed the overthrow of Gómez. As the commander of the Special Service Squadron, W. C. Cole, remarked during the crisis: "Visits of courtesy made by vessels of the Navy of the United States would have a tendency to strengthen President Gómez and the party in power."[53]

Secretary Hughes never recorded his reasons for readying the Special Service Squadron for Venezuela. Hughes and his predecessors had justified their interventions in Caribbean politics as intended either to keep European nations from meddling in the Western Hemisphere, to promote democracy, or to protect U.S. lives and property.[54] Hughes, however, could not use those justifications in supporting Gómez. In the 1920s, the war-weakened European nations had little influence in the Caribbean. And an endorsement of Gómez would help perpetuate tyranny, not establish democracy. Moreover, Cook never warned Hughes that a new Venezuelan government would harm the U.S. oil industry. In fact, Cook never told Hughes the identity, character, motives, or plans of the presumed revolutionaries. A naval demonstration, however, might help stabilize a government that was solicitous of U.S. businessmen and had helped the United States avert the energy crisis.

There never again was a need for the United States "to strengthen" President Gómez. In December 1924, the one Venezuelan capable of inspiring an anti-Gómez movement, Cipriano Castro, died in exile. With his death, Castro's followers ceased their plottings and pledged their allegiance to Gómez. Except for occasional disturbances, particularly a rebellion led by students at the Universidad Central de Caracas in 1928, Gómez ruled unchallenged until his death in 1935. In March 1927, for example, a confident and secure Gómez released political prisoners from his dungeons. Politically, Venezuela became tranquil, even lethargic. As British Minister Edward Keeling bemusedly told the Foreign Office in 1933: "There can be no country in the world which changes as little as Venezuela, and except for some obvious modifications, Sir William Seed's comprehensive report for 1925 is as applicable today as it was then."[55]

Venezuelan-U.S. relations became as stable as Venezuela's political life. The Special Service Squadron's visit of September 1923 and the Stinnes–Compañía Venezolana scare of 1924 were the last noteworthy incidents until Gómez' death. Relations between the two countries lapsed into routine cordiality, even mutual admiration. Gómez, for example, promoted good relations by allegedly bribing U.S. officials in Venezuela to speak favorably of him to Washington.[56] In addition, he was one of the few Latin American rulers who supported U.S. interventions in the Caribbean. As he told a U.S. journalist, "he had observed that the United States never intervened in a country which was governed." He even changed his fashion habits. During the war, Gómez dressed and looked remarkably like Kaiser Wilhelm II. But, after Germany's defeat, he exchanged the garb of one imperialist for the attire of another and modeled himself after his original benefactor, Theodore Roosevelt.[57] In view of this obsequious behavior, U.S. representatives in Venezuela rarely felt compelled, from the mid-1920s on, to file reports with the State Department. When they wrote, they noted that Gómez maintained order and stability and that he protected and promoted the foreign oil industry. There was little to make Washington officials anxious.[58]

Venezuelans have charged that this official U.S. friendliness and even collusion with Gómez combined with the influx of U.S. money helped the dictator stay in power.[59] The question of U.S. political and economic influence in Venezuela is a passionate one, for Gómez was a brutal and corrupt dictator, who handed over his nation's economic destiny to foreigners. During his twenty-seven–year tyranny, he crushed political dissent. The tales are legion of the atrocities which he and his henchmen committed against their opposition. Aghast at the dictator's sadism, foreign observers used historical analogies to convey the flavor of Gómez' Venezuela. British Minister Andrew Bennett noted that "the President is an 'absolute Monarch' in the most medieval sense of the word." And U.S. Chargé d'Affaires John Campbell White agreed that a "medieval severity in the treatment of prisoners" characterized Venezuela.[60]

Gómez also robbed his countrymen. Through his leasing of the national domain to the foreign oil industry and seizure of the cattle industry, he amassed a personal fortune estimated at $200 million and landholdings of 20 million acres. And he remembered his family and friends. When he was not pilfering the treasury, the tyrant was using government revenue, which had trebled during the 1920s because of the oil bonanza, to tighten his grip on the nation. He outfitted and modernized the army. With his efficient military machine, Gómez maintained internal order and discouraged dissent. In addition, he directed revenues into a massive public works program, particularly road building. Such construction gave jobs to Venezuela's sizable number of unemployed people. But Gómez, like a Roman emperor, intended not only to mollify the masses but also to use the public works program to enhance his power. Since the new roads ran from the capital to the outer states, Gómez' modernized army could move rapidly to any spot to suppress potential opposition. Thus, the dictator eradicated the regional *caudillism* that had plagued Venezuela during the nineteenth century.[61]

Gómez also used his oil income to refurbish Venezuela's international reputation. In 1930, on the anniversary of the death of the South American liberator, Simón Bolívar, he made the last payment on the foreign debt that Castro and other Venezuelan rulers had incurred. During his rule he had retired a debt of over Bs. 135 million and had completely erased the Castro legacy. As the former head of the State Department's Division of Latin American Affairs and new general manager of Gulf Oil's Venezuelan holdings, Jordan Stabler, remarked in September 1927, "the financial situation of the Government of Venezuela as regards its foreign obligations is by far the best of any other Latin American country not directly under the financial direction of the United States."[62]

Even after Gómez, his sycophants, and foreign creditors were satisfied, enough of the new oil income remained to make a significant impact on Venezuela's quiescent economy. By 1929, the foreign oil industry employed 20,000 Venezuelans in semiskilled and unskilled positions. The oilworkers received free medical treatment, housing, and a wage two to three times higher than that paid elsewhere. The oil industry also indirectly employed thousands of others. Venezuelans flocked to "boom towns" near the oil camps, offering goods and services to foreign and native oilworkers. And the oil income underwrote Gómez' public works program.[63]

The oil boom was, however, a mixed blessing for Venezuelans, for it disoriented their economy and society. The sudden surge of oil money fired inflation; for example, living costs in Maracaibo, the center of oil activity, doubled in the 1920s, with the cost of housing reportedly rising 900 percent. Caracas came to have as high a cost of living as any capital city in the world. Moreover, the development of the oil industry dislocated the already shaky agricultural sector of the economy, as it contributed to an exodus from the countryside. Peasants abandoned their subsistence farms for work in the oilfields. Maracaibo's population shot from 40,000 to 80,000. Smaller municipalities, such as Cabimas and Lagunillas, had a tenfold increase in population. The production of food, which had already been disrupted by the prewar coffee boom, continued to decline, and by the late 1920s, Venezuela began to import substantial quantities of food.[64]

Periodic food shortages combined with inflation posed severe problems for most Venezuelans. Only those who shared in the oil boom could adjust to the higher prices. The oil industry, a capital-intensive industry, employed relatively few people. Twenty-thousand Venezuelan oilworkers represented less than 1 percent of the population and perhaps 3 percent of the labor force. Urban laborers and peasants had to contend with higher prices without having any opportunity for higher wages.[65]

Venezuela became a country of contradictions. Though still basically an agricultural nation, it imported food. The government had no foreign obligations, a negligible internal debt, a balanced budget, and a sound currency; yet, most of its people lived in poverty, earning less than $100 a year. And though it was a sovereign nation, Venezuela relied on a foreign-owned and foreign-managed industry to provide it with over 75 percent of its exports and one-half of its governmental revenues.[66]

A government sensitive to such radical socioeconomic changes conceivably could have mitigated their harmful effects by using the oil income to finance an increase in agricultural productivity and the establishment of small industries. In addition, it could have required the prosperous foreign oil industry to contribute to the nation's welfare. But, beyond his immediate circle, Gómez cared more for foreigners than for Venezuelans. He allowed oilmen to dictate terms, like the oil law of 1922, to the government. And he catered to their needs. When they needed assistance, company managers pilgrimaged to Maracay, the presidential retreat, for an audience with the dictator. The Venezuelan usually expressed appreciation for the industry's efforts by granting the managers the favors they requested. In 1926, for example, Standard Oil of New Jersey asked for protection from the Motilón Indians. The Motilón had harassed engineering and geological parties working in the Perijá district, southwest of Lake Maracaibo; they resented the cutting of roads and clearing of jungles in traditional hunting grounds. They had attacked oilmen, killing several with arrows. Gómez assigned a small army to protect Standard's employees, and oil prospecting resumed in the Perijá district.[67]

The foreign oil industry, U.S. and British-Dutch, supported Gómez. It allegedly paid the appropriate bribes for the embezzled concessions, and, with its oil production, it generated new tax revenues for the dictator. Yet, the industry's money, legal and illegal, was probably not crucial to Gómez' survival. He had ruled more than a decade before oil became a significant part of Venezuela's economy, and during that period, he kept Venezuela fiscally solvent and paid one-fifth of Venezuela's international debt. Indeed, foreigners invested in Venezuela because Gómez guaranteed political order and financial integrity. U.S. Consul Harry J. Anslinger repeated the opinions of other diplomatic representatives when he wrote in January 1924: "The political stability which has been in evidence over a number of years has been a great factor towards attracting a great amount of foreign capital into Venezuela."[68]

Gómez ruled without the continual need for U.S. diplomatic support. He received his most significant aid during the first year of his rule, when the United States Navy prevented Castro from returning to Venezuela, but Gómez was hardly a creation of the United States. He was a cunning politician and experienced military leader who used his position as chief of Castro's army to construct a network of alliances with Venezuela's regional *caudillos*. Having established his government, Gómez rarely required further U.S. aid. Indeed, he maintained his rule during World War I, when the Wilson administration disdained him. After the onset of the oil boom, Venezuelan exiles may have hesitated to invade Venezuela for fear of U.S. reaction. Nonetheless, any anti-Gómez group needed military and popular support to succeed. Of the exiles, only Castro had a popular following, and none had an army to challenge Gómez. When exiles attempted two coastal invasions in 1929 and 1931, Gómez' forces routed them. Venezuelans seemed incapable of toppling the dictator. In 1934, U.S. Minister George Summerlin would report that many prominent, upper-class Venezuelans had been anti-Gómez for years but were resigned to waiting for him to die.[69]

The United States did not invade Venezuela during the Gómez years, because the dictator maintained a stable rule, deferred to U.S. leadership, and protected U.S. investments. Perhaps what is most remarkable about U.S. involvement in Venezuela is not the support the Department of State gave Venezuela, but rather its reluctance to analyze Venezuelan-U.S. relations. U.S. capital and technology flowed into Venezuela, yet few officials bothered to measure or speculate about the effects of such developments. Key questions were not asked: Did the oil industry stimulate or disrupt Venezuela's economy? Did U.S. businessmen behave honorably and in the best interests of the United States while in Venezuela? Would Venezuelans denounce the foreign oil industry for its alliance with the dictator? In effect, once Gómez died, would Venezuelan-U.S. relations become as tumultuous as those with Mexico after the fall of Porfirio Díaz?

To be sure, rigorous analysis of Venezuelan-U.S. relations depended upon accurate information, and during the 1920s, Washington mainly received glowing reports about Venezuela from its field officers. But other evidence was available. In 1925, Minister Cook admitted that "Venezuelans who have no concessions and who are not interested in shipping or in the furnishing of supplies to the oil companies, rather resent the turning over of the oil deposits of the country to the foreigners." And President-elect Hoover was warned by a friend who did business in Venezuela that "the people of Venezuela hold the United States responsible in the main for the protracted rule of Gómez, who has lost no opportunity to emphasize his enjoyment of the friendship of Washington." The only investigation the State Department conducted of Gómez and the oil industry, however, was to respond in 1926 to reports that U.S. oilworkers frequently became drunk and abused Venezuelans. The department wrote to seven companies and received assurances from them that they would correct the problem.[70]

U.S. officials probably did not ask questions about Venezuela because they assumed that U.S investments in productive enterprises, such as oil, could have only positive effects for the host nation. Such influential figures as Julius Klein, Herbert Hoover, and Charles Evan Hughes firmly believed that foreign capital and technology would promote Latin American economic growth and thereby foster political stability and financial responsibility in a chronically unstable area of the world.[71] But they, too, did not ask critical questions. Whom would political stability and financial responsibility benefit and for what purposes? The Venezuelan who benefited most from foreign investment was Juan Vicente Gómez. This failure to ask questions did not inhibit Venezuelan-U.S. relations as long as the dictator survived. When Gómez died, however, U.S. diplomats would confront issues their predecessors had ignored.

3. The Good Neighbor Policy, 1930–1939

The 1930s witnessed new directions in domestic and international policies for both the United States and Venezuela. Free of the Gómez dictatorship, Venezuelans gradually liberalized their political life and instituted economic planning at the national level. The United States, too, expanded the role of government as it grappled with the problems of the worldwide economic depression. This attack on the depression included the promotion of trade with Latin American nations coupled with the promise to befriend and respect them. The new "good neighbor" attitude of the United States proved vital to its relations with Venezuela, for Venezuelans now sharply questioned the past practices and future standing of the foreign oil industry. After having overlooked for nearly two decades the course of U.S. oil money in Venezuela, the United States was forced to consider new policies that would protect its citizens' investments and satisfy Venezuelan aspirations.

On December 19, 1935, Venezuelans learned that Juan Vicente Gómez had died. They celebrated the end of the long tyranny with a spontaneous popular outburst. Mobs marched through Caracas and Maracaibo. They burned and looted and took particular care to destroy reminders of Gómez—his portraits, statues, and buildings. And when they could apprehend them, they massacred the dictator's sycophants. In a week-long rampage, they vented twenty-seven years of frustration.

General Eleazar López Contreras, the minister of war and navy, assumed control of the government. López Contreras, like Castro and Gómez, was from Táchira. In his youth he joined Castro's army and participated in the capture of Caracas in 1899. He gradually advanced through the military hierarchy and became one of Gómez' trusted aides. He helped crush the student rebellion of 1928, the last significant political challenge to Gómez. López Contreras, however, was unlike most of Gómez' followers. An educated man, he was a historian of his nation's military past, publishing several studies. Moreover, he seemed honest. As Venezuela's representative to the 1929 Tacna-Arica arbitration, a Chilean-Peruvian boundary dispute, he received Bs. 30,000 for expenses. When he returned, López Contreras gave back one-half of this grant to the government.[1] In Gómez' Venezuela, such integrity was indeed remarkable.

Though he firmly controlled the army, López Contreras chose not to perpetuate Gómez' tyranny. He allowed the mobs to expend their fury against the symbols of the old order. He released political prisoners, allowed political exiles to return home, and lifted government censorship of the press. In February 1936, he also announced a broad program of social and economic reform.

The reasons why López Contreras espoused political and economic reform in early 1936 are not certain. The general claimed in his memoirs that he wanted to introduce political freedom. He may have encouraged reform in order to outmaneuver the *gomecistas*, who wanted another dictatorship. With the population aroused, a return to the old ways of Gómez would be unthinkable and impossible. Still, López Contreras may have been unprepared for the popular reaction to Gómez' death. For example, in February 1936, an estimated 25,000 Venezuelans withstood a fusillade of bullets, marched on the presidential palace, and demanded freedom of the press. López Contreras may have promised political freedom and economic reform in order to mollify the crowds, avoid a violent upheaval, and gain time.[2]

Although the new leader may have been surprised by the anti-Gómez reaction, the fury was predictable. During the last years of the regime, the urban middle and upper classes had become increasingly dissatisfied with Gómez. The oil boom of the 1920s had stimulated the manufacturing and service areas of the Venezuelan economy, expanding the number of white-collar and skilled workers, merchants, and industrialists. Yet, these middle-sector groups, perhaps 10 percent of the population, were frustrated, unable to gain a share of political power commensurate with their social and economic position, for Gómez trusted only his military allies and *hacendados*. They also resented having to pay blackmail money to the "King's Friends" to protect their businesses.[3] And when the global economic depression hit Venezuela, they knew that Gómez could no longer guarantee prosperity. In the early 1930s, the prices of petroleum and coffee, Venezuela's major exports, fell more than 50 percent from their 1929 levels. Government revenues declined by 25 percent. In order to balance the budget, Gómez slashed public works programs, fired some government workers, and reduced the salaries of others.[4] Prominent Venezuelans remained apolitical during Gómez' last years, for it seemed futile to contest the dictator's army. But their grievances accumulated.

One group had challenged Gómez. In February 1928, students at the Universidad Central in Caracas used a scheduled week of social events to denounce the dictatorship. Although he closed the Universidad Central in 1912 after students demonstrated against his regime, Gómez had never really won control over it. In November 1918, students, angry over the continued closing of the university, marched in celebration over the fall of Gómez' erstwhile idol, Kaiser Wilhelm. When he reopened the school in 1923, students quietly resumed their criticism. A confident and secure Gómez now tolerated the student mutterings, for he believed the "revolutionary spirit completely exterminated, for it is an anchronism in this epoch of progress and well-being."[5]

Gómez and his advisors misjudged the students' fervor. Children of the expanded middle sectors, they wanted the end of personalist rule, respect for the constitution, and broader political participation in Venezuela. On the first day of the student week, February 6, 1928, at the laying of a floral offering to Simón Bolívar, a law student, Jóvito Villalba, compared Gómez unfavorably to the South American liberator. At the coronation of a student queen, poet Pío Tamayo read verses ridiculing the dictator. Another law student, Rómulo Betancourt, later to be president of Venezuela, aroused a university audience with a passionate call for political freedom in the country. By the end of the week, the government had jailed Betancourt, Tamayo, Villalba, and another student leader, Guillermo Prince Lara, on charges of subversion. The arrests infuriated university students. Hundreds demonstrated, daring the authorities to arrest them, in Caracas' streets. Gómez responded to the disorder by ordering the arrest of 220 members of a student organization, the Federación de Estudiante de Venezuela.[6]

What followed surprised even the most romantic of the student radicals. Citizens in Caracas, La Guaira, and Maracaibo openly protested the arrests. Confused and unsettled by the popular disturbances, Gómez released the student agitators. Once free, the seemingly triumphant students resumed their protests. Young officers and cadets at the Military College soon joined them. The military men resented being used by Gómez for nonmilitary activities, such as tending the dictator's cattle. On April 7, 1928, the students and young officers rebelled against the government. They seized the unoccupied presidential palace and persuaded its defenders to join their ranks, but their success was short-lived. Loyal forces, led by General López Contreras, smashed their attempt to reach the barracks at San Carlos, Caracas' major arsenal. The rebellion was over. Gómez imprisoned dissident military officers and assigned rebellious students to road-construction crews in Venezuela's interior. As he explained to his news organ, *El Nuevo Diario,* "As they did not want to study, I am teaching them to work."[7]

A few of the students, notably Betancourt, Villalba, and Raúl Leoni, another future president of Venezuela, escaped the repression and went into exile in various Latin American countries. During the next seven years, the exiles corresponded and exchanged solutions to Venezuela's problems. Borrowing from such disparate programs, philosophies, and events as the New Deal, the Mexican Revolution, Marxism, and the Peruvian *aprista* movement, they prepared "a national revolutionary message" of political democracy, economic nationalism, land reform, and social action for Venezuela. But until Gómez died, the "Generation of '28" had to postpone action and be content with refining its ideas.[8]

During the turbulent months of early 1936, the repatriated ex-students led the anti-Gómez reaction. Villalba headed the crowd that marched on the presidential palce. Betancourt and Leoni organized the Movimiento de Organización Venezolana, a popular front movement determined "to prevent any reestablishment of despotism in Venezuela." And other former student activists founded Venezuela's first communist party, the Partido Republicana Progresista.[9]

While the Generation of '28 organized, López Contreras quietly consolidated his rule. In April 1936, the *gomecista* Congress named him constitutional president. Secure in his power, the president moved to blunt the popular fervor. In reaction to the February disorders, Congress enacted and López Contreras signed the Ley Lara, a measure designed to restrict the right of assembly. The ex-students denounced the law and planned a general protest strike for Caracas on June 9. They demanded repeal of Ley Lara, dissolution of Congress, and scheduling of general elections.

In planning the strike, the Generation of '28 received the support of the Unión Nacional Republicana (UNR), a new organization. Representing prominent Venezuelan businessmen who had cautiously opposed Gómez during the dictator's last years, the UNR supported López Contreras for president. Its members favored, however, the broadening of basic political rights, such as freedom of speech and right of assembly. The alliance between the ex-students and the businessmen soon dissolved. The coalition strike committee originally planned a twenty-four–hour strike for metropolitan Caracas. Later encouraged by their success in Caracas, the committee decided to prolong and widen the strike, and protests rocked Venezuela's major urban areas. The strike committee had made a tactical error: they failed to force concessions from the government, and the strike gradually collapsed. Moreover, the strike had frightened middle- and upper-class opponents of Ley Lara and left the ex-students isolated in their opposition to López Contreras.[10] During the next few months, the position of the Generation of '28 deteriorated. López Contreras broke their December 1936 attempt to lead Venezuelan oilworkers in a strike against their foreign employers, and then, in March 1937, he exiled the leaders of the Generation of '28.[11]

After the expulsion of the dissidents in March 1937, López Contreras ruled until the end of his term in 1941 without serious opposition. Aside from the threat of force, dissent was discouraged by using literacy, property, and tax qualifications to limit suffrage to 5 percent of the population. But López Contreras' aim was not *continuismo*. He served only five years as president instead of the traditional seven years. He also disappointed the cronies of the departed Gómez by retiring from the military and ruling as a civilian. Moreover, he enjoyed a peaceful rule by capitalizing on a national issue. López Contreras became a spokesman for the popular crusade to regulate the foreign oil industry.

The oil industry lost its shield when Gómez died. Venezuelans of most political persuasions furiously denounced the industry for having been corruptly allied with the dictator and for having defrauded the national treasury of millions of bolívares in taxes. It was a common assumption that the oil companies, particularly Standard Oil of New Jersey, had bribed Gómez and his officials to obtain concessions. Moreover, prominent Venezuelans knew that the oil industry exaggerated its deductions when it paid taxes. The oil industry paid the government in taxes a percentage of the market price of oil minus transportation costs. On their tax statements between 1927 and 1930, for examples, Standard of Indiana and Gulf claimed that it cost 68¢ to transport a barrel of Maracaibo crude to the Atlantic Coast markets. Yet, in data supplied to the United States Tariff Commission, the two companies listed their transportation costs as 33¢ a barrel. A Venezuelan government official calculated that the two companies defrauded the treasury of Bs. 56,000,000 (over $10 million) in the three years.[12] Gómez overlooked the chicanery, probably reasoning that it would be ungrateful to harry an industry that had made him one of the wealthiest men in South America. But other Venezuelans could not condone the industry's frauds. As the conservatively inclined newspaper *El Heraldo* editorialized in early 1936, "a revision must be made of our petroleum industry, which is blacker than petroleum itself."[13]

Though Venezuelans readily castigated the foreign oil industry, the government did not attempt to punish or reform the industry in 1936. López Contreras concentrated instead on solidifying his power, and no one in Congress had time to prepare a comprehensive plan for regulating the industry. In 1936, Congress made only minor adjustments in Gómez' flimsy petroleum laws. Its only legislation affecting the foreign oil industry legalized labor unions.

Even though the López Contreras administration did not enact new legislation applying to them, foreign oil executives grew wary of the new political climate. In the turbulent months after Gómez' death, they urged López Contreras to restore a Gómez-like stability to Venezuela. They argued that chaotic political conditions jeopardized future foreign investment in Venezuela. And they told government officials that student militants represented, what one oil executive called, "a far-reaching communist plot to create disturbances in Venezuela."[14] When López Contreras seemed hesitant in employing stern measures against the demonstrators, oil officials openly scorned him. Standard Oil of New Jersey trouble-shooter Major Thomas R. Armstrong boasted of the ability to overthrow López Contreras. In August 1936, another Standard executive, Leon Booker, sent a wreath bearing his company's name to Gómez' tomb. Booker intended his gesture to be a tribute to a Venezuelan ruler, who, unlike López Contreras, had appreciated the oil industry's contributions.[15]

For the foreign oil industry, the López Contreras regime, and antigovernment groups, a critical period began with the onset of the oilworkers' strike in December 1936. After the collapse of the general strike in June 1936, the remaining antigovernment groups reorganized into a single opposition party, the Partido Democrático Nacional (PDN). The PDN members never prepared a party platform, but they generally favored political and economic democracy for Venezuela. This party popularly called itself the Partido Unico de la Izquierda, the single party of the left. As with previous opposition groups, members of the Generation of '28 had key posts in the party. In the fall of 1936 the PDN began planning to participate in the municipal and state elections scheduled for January. But a now-confident López Contreras refused to legalize the party and was ready to confront his adversaries.[16]

The clash began in the Maracaibo oilfields. Encouraged and supported by some members of the PDN, twenty thousand oilworkers formed into now-legal unions and went on strike against the U.S. and British-Dutch oil industry on December 14, 1936. The workers demanded improved housing facilities, salary equity between Venezuelans and foreigners, and a raise from Bs. 7 to Bs. 10 in the minimum daily wage. While better paid than most Venezuelans, the oilworkers earned less than $3 a day, which, in the words of a North American business analyst who toured the oil camps, barely sustained life "in terms of civilized values." The industry refused to negotiate or acknowledge the new unions. The strike lasted until January 22, 1937, when López Contreras intervened. He settled the strike by imposing a one-bolívar-a-day wage increase and decreeing a compulsory return to work. López Contreras justified his intervention by alleging that PDN agitators were manipulating the strike for their own political gain. Having successfully wielded his power during the strike, the president silenced his critics. In early February, he jailed some and nullified their electoral victories. Members of the Generation of '28 had won congressional seats in the January 28 election. Finally, on March 13, 1937, he exiled for one year forty-seven of his opponents, including Betancourt, Leoni, and Villalba, for allegedly "being affiliated with communist doctrine and for activities prejudicial to public order."[17]

Oil executives gleefully watched López Contreras' campaign against the Generation of '28. They had dismissed the December strike as "communist inspired" and supplied the government with lists of alleged radicals and Communists. When López Contreras ended the strike, they noted that he said nothing about the recognition of the unions. Moreover, they sensed that López Contreras finally appreciated their place in the Venezuelan economy. Though foreign employees worked overtime, oil production dropped 40 percent during the strike. Since oil taxes accounted for over half of the nation's tax revenues, a prolonged strike would have bankrupted the treasury. On the night of March 13, oil executives gathered at the home of Alejandro Pietri, the Venezuelan attorney for Standard Oil of New Jersey, to celebrate the exile of the forty-seven "radicals and communists" and the end of political freedom in Venezuela.[18]

After March 1937, the U.S. and British-Dutch oil companies changed their policies toward López Contreras. They publicly praised his government and Venezuela's political calm. They also promised to respond positively to some of the workers' grievances. They pledged, for example, to teach Venezuelans technical skills to prepare them for partial replacement of foreigners in higher-paying positions. And they aided López Contreras' modernization program by constructing roads and a hospital in return for tax abatements. Oil officials now believed that a few conciliatory gestures would make it easier for the president to preserve their interests.[19]

As had the relationship between the South American nation and the oil industry, U.S. diplomacy toward Venezuela began to change after the death of Gómez. During the dictator's last years, the United States had, aside from diplomatic niceties, minimal official contact with Venezuela. In reviewing inter-American relations between 1929 and 1933, the Division of Latin American Affairs in the Department of State was unable to cite any noteworthy incident in Venezuelan-U.S. relations. Indeed, the highlight came when Venezuela was presented with a statue of one of the first PanAmericanists, Henry Clay.[20] No issues disrupted relations between the two countries, because Venezuela was relatively unaffected by the depression-induced turmoil that racked South America and strained inter-American relations. Unlike its neighbors, Venezuela maintained its foreign exchange at parity, imposed no exchange restrictions, and, since it was the only Latin American nation without an external debt, did not default on foreign loans. The Venezuelan leader continued to back the United States on international questions, and the U.S. oil industry, now firmly established in Venezuela, preferred to negotiate directly with Gómez and his friends and without the assistance of the legation. President Herbert Hoover's representatives in Venezuela, Minister George Summerlin, a career diplomat, and First Secretary Warden McKee Wilson, confined their activities to submitting favorable, but generally innocuous, accounts of Gómez' rule to the State Department.[21]

Venezuelan-U.S. relations remained quiet during President Franklin D. Roosevelt's first years in office. Roosevelt dedicated the United States to the role of the "Good Neighbor" in the Western Hemisphere. He pledged that the United States would treat its southern neighbors with understanding, friendship, confidence, and respect. Roosevelt repeated what most other presidents proclaimed; his contribution to inter-American relations was to give some meaning to those words by renouncing an old practice. At the Seventh International Conference of American States held in Montevideo, Uruguay, in late 1933, Secretary of State Cordell Hull announced to the pleased Latin American delegates that the United States no longer claimed the right to intervene with military forces in their countries. The administration's firm disavowal of Theodore Roosevelt's interpretation of the Monroe Doctrine removed a major source of bitterness between the United States and Latin America.[22] The

Good Neighbor policy did not, however, measurably affect relations with the Gómez regime. The Venezuelan president was one of the few Latin Americans who had supported U.S. military rule in the Caribbean.

Through 1934, the State Department kept Summerlin as minister to Venezuela. During those two years, the department had not given Summerlin any new written instructions. The department may have been subtly trying, however, to keep its distance from Gómez. British Minister Edward Keeling reported, for example, that during the twenty-fifth anniversary celebration of Gómez' rise to power, "the United States and the Latin American republics have remained rather in the background in connection with the festivities, presumably in order to show their lack of sympathy with the present political regime in this country."[23]

In January 1935, Washington replaced Summerlin with Meredith Nicholson. Nicholson was a prominent essayist and novelist as well as a stalwart Democrat from Indiana who frequently lectured for his party. He had declined an earlier offer from President Woodrow Wilson to be ambassador to Spain. During the 1920s, he worked in Indiana politics and helped manage Alfred Smith's presidential campaign. At the age of sixty-six, he accepted President Roosevelt's invitation in 1933 to be minister to Paraguay. There he helped settle the Chaco War. Since the State Department periodically rotated its representatives in Latin America in the Roosevelt administration, Nicholson's transfer to Venezuela was probably a normal procedure.[24]

During his first year in Caracas, Nicholson continued Summerlin's practice of only occasionally submitting reports to Washington. After Gómez died, however, Nicholson, true to his literary background, supplied the department with a steady stream of colorful analyses. The U.S. diplomat enthusiastically welcomed the Venezuelan political developments of early 1936. He believed that President López Contreras "sincerely intended" to liberalize the political system and establish "an enlightened form of government." He argued that López Contreras deserved the support of the United States and, in particular, the business community. When oil executives began to scorn López Contreras, Nicholson lashed out at their policies.[25]

In his despatches, Nicholson acknowledged that some Venezuelans unfairly criticized the foreign oil industry. He disputed charges that the industry provided the oilworkers with miserable living and working conditions. But he also emphasized that the foreign oilmen "are now reaping the fruits of what they have so misguidedly sown."

By bribing Gómez and openly praising him, the oil industry had thoroughly identified itself "with a regime which is now anathema to the Venezuelan people and to their responsible officials." Instead of preparing for the changes that might take place after Gómez died, the oil industry had consistently followed "a policy of shortsightedness verging on stupidity."[26]

With Gómez dead, the oil companies still refused to change. In Nicholson's opinion, the oil companies would have to transfer their chief representatives in Venezuela before restoring harmony between government and foreign enterprise. He argued that oil executives of the Gómez period, such as Leon Booker of Standard Oil of New Jersey, William T. S. Doyle of Shell, and Jordan H. Stabler of Gulf, belonged to the "old school of imperialists" who believed that the Marines "ought logically to follow American investments in foreign countries wherever required by the interests involved." In addition, the oil executives' personal lives offended Venezuelans. Booker had an illegitimate son; Doyle lived openly in Caracas with a "Spanish circus performer"; and Stabler, who had "assumptions of social and intellectual superiority," revealed a "congenital snobbishness" with his "display of a monocle only when European visitors arrive and his lonely devotion to polo in a community that looks upon a predilection for that sport as an advertisement for mental weakness." Finally, Nicholson feared the continued presence of these veteran oilmen would harm Venezuelan-U.S. relations. Doyle and Stabler were no less than former chiefs of the State Department's Division of Latin American Affairs. Venezuelans might therefore identify the oil companies' policies with U.S. foreign policy. As Nicholson wrote in September 1936, "by the nature of their interests here, these men are the most prominent representatives of the United States in Venezuela, and because of what they stand for in the Venezuelan mind, they are not making friends for their country."[27]

Both Nicholson and the oil executives misinterpreted López Contreras' initial leniency toward the antigovernment groups. His January 1937 termination of the oilworkers' strike satisfied the foreigners and disproved Nicholson's contention that oilmen who had connived with Gómez would be unable to work with López Contreras. Perhaps surprised both by the government's harsh policies of early 1937 and the oil companies' new conciliatory attitude toward López Contreras, Minister Nicholson curbed his criticism of the oil executives.

In 1936, the State Department reacted cautiously to Nicholson's analyses of the U.S. oil industry's problems in Venezuela. Laurence Duggan, the chief of the Latin American Division, described Nicholson's reports about the oil executives as "very shrewd estimates of the individuals concerned and of their attitudes" and passed them to his superior, Under Secretary of State Sumner Welles. Duggan hoped that the oil companies, in their own interest, would transfer their representatives, and he suggested that the department approach the companies. Welles rejected Duggan's suggestion. In September 1936, without providing an explanation, he informed Duggan that "under the present circumstances," he considered it "undesirable" to speak with the companies.[28]

During this contentious period between December 1935 and March 1937, the Roosevelt administration generally avoided the controversy that surrounded the Venezuelan oil industry. Yet, even by its inaction, the State Department had altered, however slightly, previous administrations' policies. Neither Duggan nor Welles rushed to rescue an industry that other diplomatic officials, such as Charles Evans Hughes and Frank B. Kellogg, had assiduously labored to establish. Moreover, the department had in Venezuela a representative who was not a fervent ally of the U.S. oil industry. Meredith Nicholson was the first U.S. official to question the industry's behavior.

Aside from routine matters, trade relations between the United States and Venezuela drew the attention of the Roosevelt administration. In June 1934, Congress enacted the Reciprocal Trade Agreements Act. The bill, which Secretary of State Hull particularly sought, empowered the president to negotiate and conclude bilateral tariff treaties with foreign countries without seeking Senate ratification. It also authorized him to cut existing duty rates by as much as 50 percent in exchange for equivalent concessions. World trade had declined precipitously after 1929, and the proponents of the bill argued that trade expansion would follow tariff reduction and help end the depression. In addition to promoting prosperity, Hull believed that unfettered foreign trade would mitigate international tensions. He argued: "Peace and Prosperity are not separate entities. To promote one is to promote the other." All nations needed "reasonable opportunities to trade with one another" in order to "sustain their people in well being." And "when nations cannot get what they need by the normal processes of trade, they will continue to resort to the use of force."[29]

From 1934 to 1936, the State Department's new Division of Trade Agreements successfully negotiated reciprocal trade agreements with fifteen foreign countries, eight of them Latin American. Secretary Hull instructed the division to concentrate on Latin America for economic and political reasons. The value of U.S.–Latin

American trade had declined approximately 75 percent during the Hoover years. Most Latin American nations had essentially one-crop economies and heavy international debts. They depended on export earnings to pay their creditors and to finance economic diversification. Accordingly, the nations welcomed U.S. initiatives to reopen a market that the Hawley-Smoot tariff of 1930 had helped close. For the United States, a revitalization of trade was also appealing. U.S.–Latin American trade was complementary. Latin American exports of raw materials and tropical products posed no competitive danger to U.S. domestic production, and, in return, the United States was able to market its manufactures and processed foods.[30]

Besides opening markets, reciprocal trade agreements would foster hemispheric friendship, cooperation, and solidarity. By renouncing intervention and reducing tariffs, Secretary Hull believed that he would eliminate the two chief sources of Latin American bitterness toward the United States. By restoring trade and prosperity, Hull also hoped to retard the spread of economic nationalism and check the expropriation of property owned by U.S. citizens. As the Latin American Division saw it, the effect of the depression had been to generate political chaos in Latin America and make the foreign investor a convenient scapegoat. As such, diplomacy in Latin America has "in the main perforce been limited to an attempt to salvage as much as possible of the American interests which have been constantly menaced by unpopularity, economic demands beyond their power to fulfill, and the clamor of nationalistic elements eager to force the foreigner to bear as large a share as possible of the onus of the financial crisis."[31]

In mid-1936, the United States and Venezuela prepared to discuss trade. Congressman Wesley Disney of Oklahoma had introduced a resolution doubling the tariff on imported petroleum. The Venezuelan Foreign Ministry, fearing the effect of the proposed tariff, expressed to Minister Nicholson interest in a reciprocal trade agreement. Secretary Hull fervently opposed any rise in the U.S. tariff, and he hoped to use a trade agreement with Venezuela to circumvent the proposed legislation. In addition, Hull and the Trade Agreements Division wanted to halt "a nationalistic trend" in Venezuela. On October 13, 1936, the Venezuelan Congress raised its tariff 2 percent on most imported goods and over 500 percent on some items, such as automobiles. As explained by Foreign Minister Esteban Gil Borges, the government needed more tax revenues and wanted to discourage the purchase of luxury items. Nicholson also opined that the Venezuelans believed a higher tariff would give them a better bargaining position for the trade negotiations. In any case, Hull both enticed and pressured the Venezuelans to cancel the tariff increases. The secretary granted Venezuela "most-favored-nation" status before trade negotiations between the two countries formally began; that is, Venezuela received the tariff concessions which the United States granted all other nations except Cuba. He also had Nicholson "vigorously and forcefully set forth the Department's point of view" that the tariff would jeopardize future trade negotiations, abet the advocates of high tariffs, and damage the international movement to liberalize trade. Having failed to dissuade the Venezuelans, the Trade Division voted on November 11, 1936, to seek an agreement with Venezuela in order "to provide safeguards against further increases in Venezuelan tariffs." The division decided, however, to delay the public announcement until Congress renewed the Trade Agreements Act for another three years.[32]

The two nations quibbled about the terms of a reciprocal trade agreement for the next two years. The major concession which the United States offered Venezuela was to reduce by one-half the duty on imported petroleum from one-half cent to one-fourth cent a gallon or 21¢ to 10.5¢ a barrel. Pressured by domestic producers and their political allies, Congress had enacted the original tariff rate in June 1932. The price of Kansas-Oklahoma crude, for example, had fallen from $1.30 a barrel in 1929 to 55¢ a barrel in 1931. The tariff had not, however, restored the domestic producers' prosperity. The problem was overproduction rather than foreign competition, coupled with the depression-induced decline in demand. Before the tariff, domestic operators produced nine barrels of petroleum for every imported barrel. The tariff and voluntary restrictions cut imports in half, but the price of domestic crude in 1933 slipped to 45¢ a barrel.[33]

The tariff had not damaged the Venezuelan economy. Before the tariff, the foreign oil companies marketed about 55 percent of Venezuela's production in the United States, approximately seventy million barrels a year. After the imposition of the tariff, their sales to the United States declined in volume by more than 50 percent.[34] The companies found new markets, however. In 1932, Standard Oil of New Jersey bought Standard of Indiana's subsidiary, Lago Petroleum. Standard of Indiana had sold all of its Venezuelan petroleum in the United States. Unlike Standard of Indiana, Standard of New Jersey had overseas markets, and it sold Lago's production in Europe.[35] The foreign oil companies maintained Venezuela's 1931 production in 1932 and 1933 and began to expand it gradually thereafter with the slow improvement of the world economy.

In trade negotiations, Venezuela professed indifference to the U.S. offer to halve the duty on imported petroleum. It held that the tariff concession would principally benefit the oil companies, for Venezuela received its royalties based on the market price of oil. Since the oil companies found outlets for the oil, the Venezuelans claimed they had no real interest in where the companies sold it. The United States countered that, despite its proclaimed indifference, Venezuela seemed worried whenever Congress debated the tariff issue. A tariff hike might further restrict U.S. imports of Venezuelan oil. Moreover, U.S. negotiators pointed out that they were offering concessions not just on petroleum but on over 90 percent of Venezuela's exports to the United States. In return, the United States asked for concessions on less than 50 percent of its exports to Venezuela, principally automobiles, light machinery, electrical equipment, and foodstuff.[36]

After prolonged haggling, the two nations struck a basic agreement in May 1938. They signed a *modus vivendi* granting each other most-favored-nation status until a treaty could be drawn and formally approved in both countries. Negotiations continued in Caracas for several months as the United States gradually overcame Venezuela's reluctance by granting some additional minor concessions. The State Department had realized that Venezuela was "lukewarm" about the treaty, for President López Contreras was reassessing foreign economic policy. By the end of 1938, negotiators had agreed on the list of concessions. The treaty became effective in November 1939.[37]

In the reciprocal trade agreement, the United States granted Venezuela double the amount of tariff concessions received. U.S. officials believed that the goals of liberal world trade policies and close inter-American relations justified generous concessions.[38] The U.S. approach to trade was not, however, entirely disinterested, for the United States was opening its market to companies owned by its citizens. Lower tariffs on petroleum profited not only Venezuela but also the stockholders of Gulf and Standard Oil of New Jersey. Furthermore, reductions in world tariff barriers would particularly benefit countries with diversified economies. If the United States convinced the twenty Latin American nations to reduce or abolish tariffs, for example, its citizens could sell manufactured goods, processed foods, and raw agricultural products to their southern neighbors. Venezuela, on the other hand, had only one major product to sell—petroleum. The other nations would be in similar trading positions, for they, too, mainly had monocultural economies. In addition, reciprocal trade agreements benefited the United States by granting most-favored-nation status to the trading partners. If either the

United States or Venezuela granted tariff concessions to a third country, the other would receive the advantages. But the United States could respond more readily to the stimulus of a new market than could a one-crop-economy nation. In essence, Hull's reciprocal trade agreements strategy was akin to Secretary of State Hughes' successful brandishing of the Open Door principle in the Middle East and Venezuela during the early 1920s. Both secretaries, who were confident of their nation's economic might, wanted to structure a world economic order that gave U.S. businessmen a chance to test their strength.[39]

In part, Venezuelan officials delayed accepting a reciprocal trade agreement, for they also knew that the agreement would ultimately favor the United States.[40] Yet, the South American nation signed the agreement for its immediate economic benefits. Through the most-favored-nation provision, Venezuela gained concessions on its two major agricultural exports—coffee and cacao. The United States bought 20 percent of Venezuela's coffee and cacao exports in 1938, and officials hoped to maintain or perhaps expand their trade. Indeed, Venezuela dramatically increased its sales of coffee and cacao to the United States. But the wartime disruption in the normal channels of trade rather than the trade agreement probably accounted for much of the trade expansion. During World War II, the United States bought virtually all of Venezuela's coffee exports and nearly 75 percent of its cacao exports. Simultaneously, U.S. exporters' share of the Venezuelan market grew from 56 percent to over 75 percent.[41] While the effects of the war exaggerated the impact of the trade agreement, the new pact, nonetheless, confirmed the patterns of trade that had been evolving since 1914. Moreover, it ensured that U.S. entrepreneurs would continue to dominate the Venezuelan market, for the trade agreement of 1939 would last until mid-1972.

In more tranquil times, Venezuela would have quickly signed a reciprocal trade agreement with the United States. The United States made tempting offers, and President López Contreras wanted amicable relations with the hemisphere's preeminent power. But in 1937 and 1938 Venezuelan officials became increasingly ambivalent about tariff concessions for petroleum, and they worried about domestic reaction to an agreement. After a temporary lull in early 1937, controversy erupted over the foreign oil industry. Venezuelans resumed denouncing the oil industry, and their legislators proposed new oil taxes. The industry's continued resistance to change further infuriated Venezuelans. For U.S. diplomatic officials, securing Venezuela's assent to a trade pact became secondary to easing the growing tension between the oil industry and the people and government of Venezuela.

The oil executives misinterpreted López Contreras' intentions when they celebrated his March 13, 1937, exile order. Though committed to the preservation of oligarchic rule, López Contreras wanted to reform Gómez' economic policies. He had assumed control over a miserably poor country with perhaps a majority of the people living virtually outside a cash economy. During the Gómez era, there was a consolidation of *latifundismo* in Venezuela; 90 percent of the peasants were landless and many were ensnared in forms of debt peonage.[42] The economic collapse of the 1930s intensified the peasants' problems. Still basically an agricultural nation, Venezuela depended on export earnings of its only cash crops—coffee and cacao—for any semblance of rural prosperity. In 1928, Venezuela earned a near record Bs. 166 million from the sale of the two commodities. Six years later, sales amounted to a disastrous Bs. 36.5 million, a decline of 78 percent. Furthermore, with the fall in world oil prices and resulting decline in royalty earnings, Gómez balanced the budget by slashing the public works program by more than 50 percent.[43] By the time the tyrant died, Venezuelan laborers and peasants were in a desperate economic position. Treasury Minister Cristóbal Mendoza found an "exhausted people" and agriculture, industry, and commerce "in a state of complete prostration." Mendoza's views were not the mere self-serving statements of a new government eager to denounce an old regime. As a British diplomat in Venezuela observed in early 1936, the Venezuelan standard of living had "been probably more wretched than that of any other supposedly civilized race."[44]

López Contreras planned to revitalize the economy with income derived from oil. The nation's new theme was "*sembrar el petróleo*," or "sow the petroleum." Gómez had squandered the petroleum bonanza while industry and commerce languished and agriculture suffered. To guarantee prosperity and economic independence, Venezuelans needed a healthy, diversified economy. López Contreras hoped to improve agriculture with "scientific farming" techniques, and he wanted to launch a new public works program to create jobs and modernize Venezuela's communication and transportation facilities.[45] During his first year of rule, he increased federal budget outlays to agriculture and health. He had some cash available, for Gómez had left a Bs. 100 million surplus in the treasury. The dictator had kept Venezuela fiscally solvent as the country's economy nearly collapsed. The surplus would pay, however, for only part of Venezuela's needs, for López Contreras talked of spending $100 million on public works. The president planned to fund his projects by collecting back taxes and imposing new ones on the foreign oil industry.[46]

López Contreras appointed Néstor Luis Pérez to head the Development (Fomento) Ministry and to prepare new regulations for the exploitation of oil. Pérez, who had survived nine years in Gómez' dungeons, first revived the charges that Lago (Standard of Indiana) and Mene Grande (Gulf) had illegally deducted Bs. 56 million in transportation charges between 1927 and 1930. In 1937, he brought lawsuits against the two U.S. oil companies. Lago decided not to contest and settled the Bs. 26 million claim out of court. Arguing that Gómez had already dismissed those charges, however, Gulf began a court challenge to Minister Pérez' Bs. 30 million claim. While that litigation dragged on, Gulf lost a different case. On April 4, 1938, the Venezuelan Supreme Court ruled that Gulf had illegally collected rebates on concessions granted in 1925 and 1926 and ordered the company to pay over Bs. 15 million in back taxes.[47]

Minister Pérez was less successful in enacting new taxes than in collecting old ones. The U.S. companies still operated under the oil law of 1922. Though the new government collected taxes more aggressively than had the Gómez regime, it garnered only about one-fourth of the companies' profits. The argument that the speculative nature of the investment in Venezuela justified generous terms for the industry seemed increasingly dubious, for oilmen had been spectacularly successful and had made Venezuela into the world's second leading oil producer behind the United States. Venezuelan wells were new and rich, and many flowed from natural pressure. Moreover, these wells were near Lake Maracaibo, and oilmen had access from the lake to the Atlantic Ocean and the world's markets. The United States Tariff Commission calculated that between 1927 and 1930 it was 46 percent cheaper to produce and ship a barrel of Venezuelan crude to a refinery in an Atlantic Coast state than to do the same with one from Texas.[48]

Pérez and the Venezuelan Congress sought a larger share of the nation's bounty with the petroleum law of 1938. The law boosted land taxes and set royalties between 15 and 16 percent. But the new legislation failed. The oil companies had forty-year leases on their concessions; they insisted on the sanctity of contract and refused to convert their concessions to the new law. Oil executives were generally confident that the Venezuelan courts would uphold their rights. Though the courts judged the industry guilty of tax fraud—something oilmen confidentially admitted—they were not necessarily swayed by political and popular pressure. For example, the Supreme

Court twice rejected challenges to a section of the 1922 oil law, the industry's right to import essential materials duty-free. Incensed by the industry's intransigence, Pérez suspended the sale of new concessions, but it was a futile gesture. The foreign oil industry already held 11 million hectares (one-eighth of Venezuela) and had exploited less than 1 percent of its holdings. Pérez' action would have the effect of limiting competition and ensuring the continued domination of Gulf, Shell, and Standard Oil of New Jersey.[49]

President López Contreras realized that the oil law of 1938 would fail. Perhaps to protect his prestige, López Contreras dismissed Minister Pérez and waited several months before signing the legislation.[50] Since the industry refused to pay the higher royalties, López Contreras curtailed his ambitious public works program. He settled for added revenues that came from collecting the oil industry's back taxes and from the industry's increased production of the late 1930s.

The 1938 clashes over taxes and royalties worried the State Department. The department approached the disputes with a different perspective and more apprehension than it had in 1936 when Minister Nicholson first warned of trouble. The question of the role of the U.S. oil company had replaced the right of intervention as the most contentious issue between the United States and Latin America. On March 13, 1937, the Bolivian government expropriated Standard Oil of New Jersey's holdings for alleged tax fraud, and almost exactly one year later, in a far more momentous and influential decision, President Lázaro Cárdenas of Mexico expropriated the properties of U.S. and British-Dutch oil companies for their refusal to accept the full terms of a decision by a Mexican labor board on a 1936 labor dispute.[51] The department, which had been surprised by and unprepared for both the Bolivian and Mexican actions, decided that it must prevent a third loss of U.S. oil properties in Latin America.

Less than three weeks after the Mexican expropriation, the Venezuelan Supreme Court ruled against Gulf on the illegal rebates claim. A harried State Department anxiously asked its representatives in Venezuela if another crisis loomed. The department's fears were premature and perhaps unrealistic. Oil dominated the Venezuelan economy; any disruption in production and sale would bankrupt the government and plunge the nation into economic chaos. Mexico, which produced one-fourth as much petroleum, had five times as many people, and a more diversified economy than Venezuela, could weather a decline in petroleum income. Even the most vociferous critics of the foreign oil industry understood that Venezuela could not afford expropriation. When the oil companies ignored the 1938 oil law, López Contreras could not counter with the trump of expropriation.[52]

In fact, the López Contreras government opposed expropriation. The president and his supporters were political and economic conservatives, with a strong commitment to capitalism. They did not have well-defined philosophical objections to foreigners controlling their economy, although they frankly admitted that, with one-third of the national income dependent on the oil business, Venezuela's destiny was "in the hands of the directing executives of a few corporations." What they resented was the foreign oil industry's past meddling in their politics, and what they wanted was a larger share of the profits from the sale of their country's natural resource, but within "business reason." That money would then be used to improve the lives of the Venezuelan poor. López Contreras stated that he wanted to negotiate fairly with the oil companies; he tried, for example, to convince the industry to settle the tax fraud cases out of court and thereby avoid antagonizing the public. When the companies insisted on their legal rights under the old oil law, the government had no other strategy for obtaining additional petroleum revenues.[53]

U.S. representatives in Venezuela analyzed López Contreras' oil policies and assured the department that he had no intention of expropriating U.S. properties.[54] Yet, some U.S. diplomats fretted. The constant wrangling between government and industry would only increase tension and perhaps allow passion to overcome reason. A Venezuelan expropriation would be another severe blow to the Good Neighbor policy and would endanger U.S. commercial and strategic interests. For some officials, it was essential that the State Department forego its passive attitudes and mediate between the Venezuelan government and oil industry.

Laurence Duggan, the chief of the Division of American Republics, led the fight for a new policy toward Venezuela. Duggan, the son of Stephen Duggan, a prominent political scientist, had joined the department in 1920, and five years later at the age of thirty he became the division's head. During his years in the department, Duggan had demonstrated a keen sensitivity to the socioeconomic aspirations of Latin America. For example, he had commended Meredith Nicholson for his astute reporting on post-Gómez Venezuela, and he had tried to convince Under Secretary Welles to speak to U.S. officials about transferring their representatives. During both the Mexican land and oil expropriation crises, he sided with Ambassador Josephus Daniels and against Welles and Hull in arguing that the "sacred character of property rights" would have to be balanced against the Mexican government's goals of improving its people's living and working conditions. In Duggan and Daniels' opinion, Latin Americans, as well as North Americans, deserved a "new deal."[55]

Throughout 1938, Duggan urged Welles and Hull to understand the Venezuelan position on the regulation of the foreign oil industry. He repeated Nicholson's views that during the Gómez period the oil industry "exercised enormous power and was in a position to secure whatever it wanted and practically on its own terms." He assured them that the Supreme Court decision against Gulf was not analogous to the Mexican expropriation. And he sent them a letter from Daniel Scott, a former State Department employee. Scott, who had been in Caracas, wrote that local oilmen believed that López Contreras wanted increased royalties, but that he had no desire to emulate the Mexicans. Yet, in Scott's opinion, "the oil officials in New York are deliberately building up the Mexican bogey with the Department for the purpose of creating a mental state favorable to a strong stand in Venezuela if real threats develop."[56]

Duggan's campaign failed to inspire a review of Venezuelan-U.S. relations. Secretary Hull decided to take a firm stance on the Mexican expropriation. While conceding the right of expropriation, Hull insisted that the oil companies must be justly compensated. In a forcefully worded note, Hull virtually demanded that President Cárdenas provide written guarantees to his oral assurances that Mexico intended to pay for the expropriated properties. Ambassador Daniels considered Hull's note an intemperate ultimatum likely to lead to an open break between Mexico and the United States. Without authorization, Daniels decided to show Mexican diplomats the note, but, in order to spare Mexican sensibilities, he chose not to deliver it "officially."[57] In any case, during 1938, Hull wanted to pursue a tough policy in the defense of U.S. property, and he was probably not susceptible to calls for a new understanding of Venezuela's relations with its oil industry. Indeed, he may have believed a stern attitude toward Mexico would discourage a Venezuelan expropriation. Speaking for Hull, economic advisor Herbert Feis told Secretary of the Treasury Henry Morgenthau: "If Mexico can make this expropriation without adequate payment, the whole American oil properties throughout Latin America would probably be fairly much jeopardized."[58]

Duggan persisted in his bureaucratic struggle. In February 1939, he visited Venezuela to assess government-industry relations, and on March 17, he submitted his findings to Welles and Hull. Duggan judged President López Contreras a moderate who was trying to chart a middle course between extremists and the *gomecista* reactionaries. He thought López Contreras would accomplish his mission, unless an impasse developed over oil. Duggan emphasized that Venezuelans fervently believed the government and the oilworkers received only the "skimmed milk" of oil profits. And he warned that the oil issue was explosive and that the "weight of events" could lead to expropriation.[59]

In arguing for a new posture toward Venezuela, Duggan was probably aided by developments in the Mexican oil case. Hull's Mexican policy fared badly in late 1938 and 1939. A variety of people had helped sabotage it. Ambassador Daniels continued to moderate diplomatic exchanges. In his conciliatory approach, Daniels seemed to have the support of his good friend and former Navy Department colleague, President Roosevelt. In a news briefing, Roosevelt indicated that the United States would never support any exaggerated claims by the oil companies and that he had confidence in President Cárdenas' pledges. Hull was further handicapped by the reluctance of other departments to back his policies. Treasury officials worried that strong diplomatic pressure would force Mexico into the Axis camp. Secretary Morgenthau refused to use the daily purchases of Mexican silver as an economic bludgeon against Mexico. As the crisis persisted, the treasury's fears seemed particularly justified. Attempting to foment economic chaos in Mexico after the expropriation, the Anglo-American oil industry organized a global boycott of Mexican oil. Its united front failed when William Rhodes Davis, a U.S. oil maverick of uncertain reputation, contracted to sell and transport Mexico's oil to Nazi Germany. Assured of a steady income and with fervent, near frenetic popular support, President Cárdenas could resist Hull's pressure and reiterate his intention to pay the oil industry what Mexico believed it deserved, not what the oil companies claimed.[60]

With Hull's Mexican policy in disarray, high State Department officials finally heeded Duggan's arguments and chose a new approach toward the oil company problem. The catalyst for change came on June 15, 1939, when Winthrop Scott, the chargé d'affaires in Caracas, submitted a sixteen-page analysis of the government-industry feud. Scott grimly reviewed the increasing pressure which the López Contreras government and leftist critics had applied to the oil companies. In his view, the companies were "being subjected to unfair attacks of one sort or another designed to force enormous concessions from them." Scott feared that "a situation might conceivably develop leading to expropriation." Accordingly, the department must inform López Contreras about its attitude toward "the spoilation of legitimate American interests." Scott asked for permission to speak with the Venezuelan president.[61]

The relentless Duggan immediately rebuffed Scott's suggestions. In a lengthy memorandum to Welles and Hull, Duggan repeated the arguments he had made in March after returning from Venezuela. He conceded Scott's contention that the oil companies suffered incessant harassment from the government and public. But he again emphasized that in the Gómez era "the companies were arbitrary, high-handed, insensitive, and ruthless" and were now "reaping the attacks of a people embittered and determined and that feels unjustly treated." He recommended that the department mediate between the industry and the government and convince the oil companies to transfer their representatives.[62]

In the first section of his memorandum, Duggan essentially restated the views that he and Nicholson had held for three years. But in mid-1939, Duggan had compelling, new arguments to add to his campaign. In the late 1930s, U.S. oil companies had discovered vast new fields in eastern Venezuela, particularly in the state of Anzoategui. Oilmen now suspected that Venezuela's "possible reserves run into astronomical figures" and that Venezuela, unlike Bolivia and Mexico, would be a significant producer for years. Moreover, these reserves might be strategically crucial. If war came, Great Britain would lose safe access to its Middle Eastern oilfields and would have to rely on Venezuela for the petroleum needed to combat the Fascist powers. In addition, Duggan pointed to the effects of the repeated conflicts between U.S. corporations and Latin American nations. He predicted that "a major clash between American oil companies and the Venezuelan government, regardless of outcome, would have a very disturbing effect on direct investments everywhere in Latin America." In order to protect these strategic and commercial interests, Duggan concluded: "I believe this government must be prepared to go further than may be customary in advising the American petroleum companies in the course they should pursue. It must not be permitted them (as occurred in the case of the Mexican dispute) to jeopardize our entire good neighbor policy through obstinancy and short-sightedness. Our national interests as a whole far outweigh those of the petroleum companies."[63]

What followed in the month after Duggan's June 26 memorandum is not clear. Perhaps Welles and Hull debated the merits of Duggan's proposals. As late as July 12, Secretary Hull seemed determined to persist in his strident defense of U.S. property. At a meeting with Diógenes Escalante, the Venezuelan ambassador to Washington, Hull lectured him on the pernicious practice "of taking things that do not belong to us and not paying for them within any reasonable time."[64] Hull referred to Mexico, but he may also have been indirectly warning Venezuela. But, in a despatch ten days later to Chargé Scott, Welles outlined a new policy. He agreed with Scott that a U.S. official should have a conversation with President López Contreras, but the newly appointed ambassador, Frank P. Corrigan, rather than Scott, would meet with the president. Corrigan would inform López Contreras that the department wanted to hear about grievances that Venezuela had against the U.S. companies. In addition, Corrigan would meet with oil company executives in New York and assure them that the department wanted to help them resolve their difficulties.[65]

The decision to adopt a new policy was a victory for Division Chief Duggan. In explaining to Scott Washington's determination to mediate between government and industry, Welles echoed Duggan's arguments. For Duggan, the persuasion of Welles was probably crucial to his triumph. Beginning in the late 1930s, Welles assumed overall control of the department's Latin American policy, for with the world order crumbling in Asia and Europe, Secretary Hull gave little attention to Latin America. It is difficult to determine which of Duggan's arguments Welles found most convincing, but his emphasis on the strategic significance of Venezuelan petroleum was probably influential. From 1939 to 1943, Welles ardently worked to ensure that Latin America supported the war effort. He helped design economic and military aid packages and multilateral defense agencies.[66] He probably found Duggan's call for a new Venezuelan policy compatible with his goal of a unified hemisphere.

The Department of State's July 1939 decision to pursue a new policy was timely and crucial. Within two months, Great Britain was at war, depending on Venezuela for as much as 80 percent of its petroleum imports. The Venezuelan government, aware of petroleum's strategic significance, faced with mounting budgetary problems, and incensed at the oil companies' obstinancy, would soon after begin a new campaign to regulate the foreign oil industry. Washington needed its new oil policy to keep peace in the oilfields and to ensure the flow of oil to Great Britain.

4. Wartime Policies, 1939–1945

The exigencies of World War II brought new challenges to the United States in its relations with Venezuela. Cognizant of its strategic worth and frightened by the prospect of it being under the sway of Nazi Germany, U.S. officials worked diligently and successfully to win Venezuela's allegiance. This effort included arranging for a new oil law that satisfied Venezuelan demands for more revenue from the foreign oil industry. In addition to securing Venezuela's support, these wartime policies had the effect of expanding the political, military, and commercial influence of the United States in Venezuela.

In trying to promote its national security between 1939 and 1941, the United States focused its foreign policy on Latin America. Officials foresaw that, if the United States became involved in the impending world conflict, they would need the political and economic support of their southern neighbors. Hemispheric solidarity would secure the southern flank of the United States and free its armed forces for other areas. Moreover, it would give the United States access to Latin America's strategically vital raw materials, such as copper, manganese, and petroleum.

U.S. officials particularly thought that cooperation with Venezuela would be critical to hemispheric defense. If enemy bomber planes were stationed in Venezuela, they could easily reach both the Panama Canal and Puerto Rico. Similarly, enemy troops could use Venezuela as a staging area for attacks on British and French possessions in the Caribbean and for the capture of the large petroleum refineries in the Dutch West Indies. And any enemy could use Venezuela's most abundant natural resource—petroleum.[1]

Realistically, the Axis powers had little hope of either occupying Venezuela or wooing it to their side. The United States had sufficient power to block an invading force, and its position on international questions was consistently supported by the López Contreras government. For example, at the prewar inter-American conferences, the Venezuelan delegation introduced resolutions that the United States wanted Latin America to adopt, such as pledges of solidarity in case of aggression from outside the hemisphere.[2] Nevertheless, the Roosevelt administration worried about Venezuela. "Fifth column" subversives might blackmail the Venezuelan government or sabotage oil production. And German entrepreneurs might disrupt U.S. trade with Venezuela.

As outlined by Propaganda Minister Joseph Goebbels in September 1933, a prime goal of Nazi foreign policy in Latin America was to win sympathy for the Reich's international policies. Goebbels directed German diplomats to organize German nationals living in Latin America into local National Socialist parties to promote the ideas of the führer and devotion to the fatherland. But, at least before 1937, the Nazis enjoyed little success in Venezuela. Count Franz von Tattenbach, the German minister to Venezuela, reported that Venezuelans were unenthusiastic about National Socialism. In particular, the Catholic church, an influential institution in Venezuelan life, lectured against the "neo-pagan" characteristics of the Nazi movement. Another handicap was that Venezuela received its international news from non-German sources, mainly North American–based wire services. Venezuelans read unflattering accounts of Adolf Hitler's rise to power and his regime's persecution of Jews. As for German nationals residing in Venezuela, Tattenbach had more gloomy news for the Foreign Office. Venezuela's Germans, who numbered between two thousand and three thousand, had assimilated into Venezuelan culture and refused to accept National Socialism's racial policies. A Venezuelan Nazi party had existed since 1926, but as late as 1936 it claimed only eighty-six members.[3]

Accounts of German activity in Venezuela after 1937 are murky and conflicting. Alton Frye, in his study of German political efforts in the Western Hemisphere, argues that "like an ominous German U-boat, Nazi influence in Latin America broke the surface in 1937." In early 1937, Dr. Edwin Poensagen replaced Tattenbach as minister to Venezuela. Tattenbach, a Weimar Republic appointee, had not promoted National Socialism in Venezuela, and he retired to Costa Rica after his replacement. It is unclear, however, whether the appointment of Poensagen, who had previously served in the South American section of the German Foreign Office and in Mexico, signaled a new Nazi interest in Venezuela.[4] Making one of its first surveys of foreign influence in February 1938, the U.S. legation in Caracas reported that it knew of no Nazi subversives. Not until 1940, when Germany had demonstrated in Norway the peril of subversion, did U.S. representatives begin to warn the State Department of the dangers of Nazi sabotage. They feared that Venezuela's Germans were tightly organized and that the Venezuelan government was not fully aware of the problem. They conceded, however, that rumors abounded in Venezuela and that they lacked evidence to substantiate their fears. In January 1941, President Roosevelt read an alarming report that noted Venezuela's Germans were fanatic Nazis and that "Venezuela is one of the few countries in Latin America where the people seem to be swinging toward Nazism." But the report's author, William S. Paley of the Columbia Broadcasting System, had not visited Venezuela. Two months later, Sir Donald Gainer, the British ambassador to Venezuela, confidently informed the Foreign Office that the fifth columnists were "very quiet" and that the Venezuelan public was generally sympathetic to Britain.[5]

Nazi agents wasted whatever efforts they made in Venezuela. The López Contreras government approved of the inter-American resolutions recommending the strict supervision of foreigners and outlawed political activity by foreigners residing in Venezuela. Indeed, López Contreras prohibited any form of political extremism, from communism to fascism. Between 1937 and 1945, no sabotage attempts were recorded in Venezuela.[6]

Nazi Germany probably placed more emphasis on developing its trade with Venezuela than for establishing political influence there. Historically, Germany was one of Venezuela's chief trading partners. Both before World War I and during the 1920s, it supplied Venezuela with between 15 and 20 percent of its imports, and the German merchant community had been, prior to the arrival of American capitalists in the 1920s, the South American nation's chief source of credit. Though both the value and the volume of its exports declined in the early 1930s, Germany's relative position in the Venezuelan market improved as it supplanted Great Britain as Venezuela's second leading source of imports, after the United States. As an importer, Germany was usually Venezuela's second or third best customer, with substantial purchases of coffee. Germans had one significant investment in Venezuela, the railroad (Gran Ferrocarril de Venezuela) running between Caracas and Valencia.[7]

Under Hitler, Germany created a barter system, known as the *aski* system, for world trade. A country that exported to Germany received for its sale a special, nonconvertible currency that could be used for purchases only in Germany. The intended effect was to eliminate multilateral trade and to give Germay the power to expand its exports to a country in exact proportion to the increase in its imports from it. With Venezuela, Germany sought to take the *aski* system one step further. The Reich's expanding military-industrial complex required increasing amounts of imported petroleum. Nearly 50 percent of Germay's petroleum imports came from Venezuelan wells. Germany did not, however, buy the petroleum from Venezuela but rather from the U.S. and British-Dutch oil companies, which shipped Venezuelan crude to refineries in Aruba and Curaçao and then sold the refined product in Europe. Since it wanted to increase its imports of petroleum, Germany made a proposition to Venezuela in 1937 that was both an offer and a threat. It proposed that Venezuela accept its oil royalties in kind (barrels of oil) instead of taxes and sell the royalty oil to Germany. The sale of oil would be directly linked to German purchases of coffee; Germany would contract for bags of coffee when Venezuela agreed to the royalty oil scheme.[8]

Germany failed to convince or to intimidate Venezuela into accepting these trade proposals. The foreign oil companies had already submitted production plans for 1937; it was impossible for Venezuela to obtain its royalties in kind. In any case, Berlin had overestimated Germany's impact on the Venezuelan market. As Minister Poensagen explained, Venezuelan officials seemed to regard Germany's coffee purchases as "trivial." They pointed to the severe imbalance of trade, with Germany selling to Venezuela twice as much as it bought from it. Poensagen warned his Foreign Office that a boycott of Venezuelan coffee would only arouse Venezuelans, imperil the Gran Ferrocarril de Venezuela, and prevent German capital from participating in President López Contreras' planned public works program. All of that would only redound to the benefit of the United States.[9]

By mid-1938, the Reich had lost any chance of gaining special access to Venezuela's petroleum. By signing the *modus vivendi* for the reciprocal trade agreement with the United States, Venezuela adopted the liberal international trade policies of Secretary of State Cordell Hull. U.S. officials knew of Germany's trade efforts in Venezuela. In fact, they were one of the reasons for granting Venezuela additional tariff concessions to ensure that it accepted a reciprocal trade agreement. Venezuelan-German trade remained at normal levels until September 1939, when it ended with the start of World War II and the British naval blockade of Germany. Germany suffered an additional blow in November 1943, when Venezuela, at the urging of the United States, nationalized the Gran Ferrocarril.[10]

Compared to its activities in Argentina, Brazil, and Uruguay, Germany seemed to undertake a minor effort to influence Venezuela to back its foreign policies.[11] German officials perhaps reasoned that, as a Caribbean nation, Venezuela was under the unshakeable sway of the United States and that it would be more fruitful to concentrate on those southern South American nations with large German populations. The Roosevelt administration perceived, in fact, the greatest danger to be in southern South America and specifically designed its prewar inter-American policies for effect there. U.S. officials employed those policies throughout the hemisphere, however, for they were uncertain of the Nazis' Latin American goals; and in the interests of inter-American harmony, they wanted to treat all the southern neighbors equally. Some of these policies helped ensure that Venezuela supported U.S. foreign policy.

One key part of U.S. prewar policy was the transfer of war matériel to the Latin American nations. With congressional approval of the Lend-Lease Act in March 1941, the Roosevelt administration gained the power to supply friendly nations, particularly Great Britain, with armaments. The legislation was a significant triumph for the president in his campaign to convince the public that the United States could not safely stand aloof from the war in Europe. In April 1941, the United States extended the lend-lease program to Latin America when Roosevelt certified that the defense of Latin America was vital to the defense of the United States. Diplomatic officials recognized that Latin Americans were anxious to see if the United States intended to fulfill pledges made at the inter-American conferences to defend the hemisphere against aggression. In addition, the State Department needed the program to woo military men in key Latin American countries. For example, the department received warnings from both its representatives in Brazil and pro-U.S. Brazilians that the cooperation of that country's armed forces was contingent upon the delivery of armaments. U.S. military men regarded Brazil's cooperation as critical since its Natal area was vulnerable to a German invasion. The State Department eventually sent over $300 million worth of matériel to placate Brazil.[12]

As with Brazil, the United States used its military aid program in Venezuela for political and diplomatic purposes. During the late 1930s, Venezuela had some military ties with Italy. The López Contreras government purchased naval vessels from Italy, sent military officers to study there, and entertained an Italian aviation mission in Venezuela. These ties did not mean that Venezuela supported Italy on European questions but simply that Italy made Venezuela inexpensive offers. In early 1939, for example, Venezuela shopped for naval equipment in the United States. But the United States lacked surplus vessels to sell, and, in any case, the Roosevelt administration opposed, on principle, the sale of war matériel to Latin

America.[13] This policy began to jeopardize hemispheric defense and prompted a reconsideration. During 1940, U.S. military officers discussed cooperative defense measures with Venezuela. In June, the discussions appeared successful, and the embassy spoke of "the magnificent cooperative spirit which Venezuela now has toward the United States." But, by August, the embassy noted that Venezuela's enthusiasm had diminished because the United States had not bolstered its defense plans with the transfer of war matériel. The lend-lease program permitted the United States to rekindle Venezuelan enthusiasm, to end Italian influence, and to secure the defense of the southern Caribbean. In July 1941, Washington promised to sell Venezuela $20 million worth of military equipment for approximately $9 million. That year, too, U.S. military advisors were assigned to Venezuela.[14]

The State Department also developed economic aid programs to attain hemispheric solidarity. In September 1940, at the urgent request of President Roosevelt, Congress increased the lending authority of the Export-Import Bank from $100 million to $700 million. As originally constituted, the bank was to increase U.S. exports by extending credits to purchasing nations. With the new funds, the bank developed financial aid programs to spur imports as well as exports. It helped underwrite, with the enlarged Reconstruction Finance Corporation, purchases of Latin American raw materials. The United States wanted to stockpile essential materials in case of war. In addition, the stockpiling program furthered inter-American cooperation by stabilizing the economies of Latin American nations which lost their European markets. The bank's second new function was to grant loans for development projects, such as the establishment of new industries. As with military aid, the State Department designed the new program particularly for Brazil. The bank granted Brazil the largest single loan, $45 million for the purchase of U.S. materials for the construction of the Volta Redonda Steel Mill.[15]

Venezuela received part of the U.S. economic aid package. The Export-Import Bank extended Venezuela money to help pay for lend-lease, small loans for agriculture and trade, and, in 1942, a $20 million loan for public works projects. The State and Treasury departments also provided Venezuela with financial advice. López Contreras wanted a review of Venezuela's fiscal and monetary practices, and he asked Washington for assistance. The United States acceded to the Venezuelan request by sending in August 1939 a mission headed by treasury official Manuel Fox to analyze the Venezuelan economy. Officials believed that, in view of the "growing importance of commercial and other relations between the United States and Venezuela," it was "especially desirable" to grant the request. Moreover, the Fox mission would ensure that Venezuela did not seek German financial advice.[16]

The Roosevelt administration's persistent courting of Venezuela helped forge a strong alliance. Two days after Japan attacked Pearl Harbor, Venezuela officially declared its solidarity with the United States and on December 31, 1941, severed relations with Germany, Italy, and Japan. During the initial stages of the war, U.S. military officials preferred that Venezuela remain neutral in order to deny the German navy any right to shell Venezuela's coast. Venezuela's neutrality was merely a convenient legal fiction, for Venezuela granted U.S. ships and airplanes special access to its ports and airstrips. In February 1945, when Germany no longer had the military potential to attack the Caribbean area, Venezuela declared war on the Axis in order to be eligible for membership in the emerging United Nations Organization.[17]

The United States similarly succeeded in convincing other Latin American nations to become military partners. By February 1942, seventeen of the other nineteen nations had either declared war on or severed relations with the Axis powers. A year later, Chile became the nineteenth Latin American nation to break relations, leaving Argentina the only "neutral" nation in the Western Hemisphere. Of those nineteen, sixteen nations sanctioned the development in their territory of air and naval bases that were available to U.S. forces. And two nations—Brazil and Mexico—actively participated in the war; Brazil sent an expeditionary force to Italy, and Mexico sent an air squadron to the Pacific.

Latin America's economic contributions to the Allied war effort overshadowed its military undertakings. Though undoubtedly useful, Latin America's military cooperation with the United States was not essential either to the defense of the hemisphere or to the defeat of the Axis. The United States Navy's crippling of the Japanese fleet at the Battle of Midway Island in June 1942 and the Anglo-American invasion of German-occupied North Africa in November 1942 removed, early in the war, the principal threats to the hemisphere's western and eastern flanks. And the participation of Brazil and Mexico in the war was more symbolic than strategically significant. Latin America's commodities and raw materials were, however, crucial to the Allied victory. During the war, nonmilitary agencies of the U.S. government bought nearly $2.4 billion worth of commodities from Latin America out of an approximate total $4.4 billion spent throughout the world. Moreover, the United States relied on Latin America for such strategically vital materials as beryllium, copper, manganese, mica, quartz crystals, tantalum, tin, tungsten, and zinc. Even recalcitrant Argentina sold its beef and wheat to the Allied nations. In effect, Latin America served as an arsenal for the United States and the United Nations. After the war, U.S. military officials agreed: "Nobody knows how much we relied on the South American and Central American countries for commodities and things that we simply had to have."[18]

Like that of other Latin American nations, Venezuela's primary role in World War II was as a supplier of raw materials. For example, in October 1942, Venezuela agreed to sell all its exportable rubber to the United States. In their rapid advance through East Asia, the Japanese had captured the world's major rubber-producing areas. Venezuela's response relieved slightly the U.S. shortage of a vital material. Venezuela's chief contribution to the war effort was, of course, not rubber, but petroleum. As Laurence Duggan had prophesied in June 1939, Great Britain would depend on Venezuela's oil for the fight against Nazi Germany. Especially during the first years of the war, Venezuelan wells supplied Britain with as much as 80 percent of its oil imports.[19]

The Department of State's decision in July 1939 to formulate a new oil policy helped ensure that Venezuela's petroleum fueled the Allied war machine. More than either blunting Axis subversion or defending the southern Caribbean, the department's chief wartime concern for Venezuela was to keep the oil wells operating. During the war, the department had to mediate between the Venezuelan government and the oil industry in order to achieve that major foreign policy goal.

In September 1939, during the first month of World War II, Dr. Frank P. Corrigan arrived in Venezuela as the new U.S. representative, replacing Minister Antonio C. Gonzalez. Gonzalez, Meredith Nicholson's successor, had resided in Caracas less than a year; he resigned his diplomatic post because of his wife's illness. In presenting his credentials to the Venezuelan government, Corrigan became the first U.S. ambassador to Venezuela. In 1939 the State Department raised the status of its mission in Venezuela from legation to embassy as a sign that it recognized Venezuela's commercial and strategic significance.

Ambassador Corrigan had an unusual background for a diplomat. He was a distinguished surgeon from Cleveland who published in medical and scientific journals. In 1934, he left his medical career to work in Roosevelt's New Deal program. The economic depression had imposed cruel burdens on medicine; Corrigan's hospital had been unable to afford medical supplies. Dr. Corrigan applauded the New Deal's social welfare policies, and he believed that he had a duty to join with others "who are trying to shape our government along the lines which unbiased scientific and technical opinion indicated it should be guided."[20] Through his friendship with the surgeon general and White House physicians and with the backing of Ohio Democrats, Corrigan won an appointment as minister to El Salvador in 1934. He also had gained some experience in Latin America when he worked in a hospital in Chile from 1917 to 1919. Corrigan served three years in El Salvador and then a year as minister to Panama before receiving his Venezuelan assignment.

On his way from Washington to Caracas in 1939, Corrigan stopped in New York to meet with officials of Gulf, Standard Oil of New Jersey, and Standard Oil of New York. He stressed to them that the State Department wanted to protect their legitimate interests and avoid a repetition of the Bolivian and Mexican crises. He asked them to take a broad view toward their problems and to understand that social conditions throughout Venezuela, not just conditions in the oilfields, would affect Venezuelans' attitudes toward the foreign oil industry. Corrigan's visit was part of the new determination in Washington to anticipate and resolve conflicts between the U.S. investor and the host country. Corrigan did not, however, approach President López Contreras about Venezuela's views on its relations with the industry. The State Department instructed Corrigan to postpone his planned meeting with López Contreras since the 1939 Venezuelan Congress had not enacted petroleum legislation that the industry considered objectionable. Nevertheless, the department cautioned Corrigan to remain alert for any potential problems.[21]

In October 1939, Corrigan submitted to Washington his first comprehensive analysis of Venezuela's oil policy. In effect, Corrigan presented the department with views that he propounded for the next eight years as ambassador to Venezuela. He judged the oil industry's local officials as hard-working, conscientious employees. Yet, Venezuelans looked upon them "with some suspicion," for they were tainted "by their very years of experience in dealing with a Government wholly different in character from the present one." Furthermore, the local oilmen antagonized politically moderate and liberal Venezuelans by maintaining their contacts with those formerly associated with the Gómez regime.[22] Corrigan agreed with Nicholson and Duggan's argument that the oil industry would have to transfer local managers before achieving peace between government and industry.

But, in Corrigan's opinion, the key issue transcended personnel questions. As he had told oil executives in New York, Venezuela's oil policy had to be considered within the context of Venezuela's socioeconomic conditions, for "when an industry becomes such a ruling factor in the life of a nation, it perforce finds itself shouldered with social problems that have little to do with business as it is usually considered." Venezuela's housing and sanitation were "in miserable condition," and the López Contreras government desperately wanted to improve the people's plight. The foreign oil industry now had the opportunity both to refurbish its image and to aid Venezuela by willingly contributing a part of its income to the government to build homes, schools, hospitals, and roads. The oil companies had to be "forward looking" and realize that they were an integral part of Venezuelan society. If nothing was done to resolve Venezuela's pressing social and economic problems, they might crystallize into political issues. Corrigan surmised that the López Contreras regime was "not now radical" or even hostile to the oil industry, but he warned that a deterioration in Venezuela's economy might force the government to emulate Bolivia and Mexico.[23]

Ambassador Corrigan's call for a socially responsible foreign oil industry produced no immediate effect. The oil industry enjoyed another peaceful year in 1940 and saw no reason to design new policies. Indeed, its position seemed particularly secure, for, in February, the Venezuelan courts had upheld once again the validity of a section of the 1922 oil code. Despite this relative calm, some officials of the Division of American Republics remained nervous. With the war underway, they feared that a disruption in Venezuela's oil production would imperil Great Britain. They regarded Ambassador Corrigan's recommendations as a way to preserve stability in the oilfields. Yet, though it recognized the need for leadership, the State Department refrained from pressing the oil companies. High officials hoped that the calm in Venezuela would endure.[24]

Though it still seemed unwilling to serve as a broker between government and industry, Washington took one step to ensure Venezuela's cooperation. In November 1939, the United States and Venezuela formally signed the reciprocal trade agreement reducing the U.S. tariff on imported petroleum by 50 percent. A month later, Secretary of the Interior Harold Ickes imposed a quota limiting petroleum imports to 5 percent of domestic production. The quota confirmed a practice that had existed since 1932 when oil companies voluntarily agreed to restrict imports in order to protect the price of domestic crude. The Trade Agreements Division used the quota to aid Venezuela and Colombia. In the five-year period before expropriation, Mexico supplied the United States with approximately 13 percent of its imported petroleum. After the expropriation and with the oil companies' boycott, Mexico's share of the market fell to 3 percent. In allocating the quota, the Trade Division based its assignments on the period of the first ten months of 1939, when Mexico's share was the lowest and Venezuela and Colombia's the highest. Venezuela received a 92.2 percent share of the market, and Colombia a 4 percent share. Mexico had to compete with other oil-producing nations for the remaining 3.8 percent.[25] As Duggan remarked to Hull, it would have been politically explosive to give an allocation to Mexico "in view of the almost unfavorable comment which such a course would occasion in the press."[26] Nevertheless, the Trade Division's action had the effect of punishing Mexico and rewarding those oil producers that respected U.S. property.

Favorable court decisions and special import allocations could not, however, protect forever the oil industry from paying a larger share of its income to the Venezuelan treasury. As Ambassador Corrigan had predicted in October 1939, a downturn in Venezuela's already shaky economy would force the government to reconsider its oil policies. With the loss of some of the European markets because of the war, Venezuela's oil production fell 10 percent from its 1939 peak. To replace the lost revenue, López Contreras began in January 1941 to press the lawsuit against Gulf for its alleged tax fraud between 1927 and 1930. The suit had lingered in the courts for four years with the government unsuccessfully attempting to persuade Gulf to settle out of court. With only four months left in his presidential term, López Contreras wanted to collect the back taxes and cap his administration with a crowd-pleasing triumph.

López Contreras' plans shot the State Department into action. Under Secretary Welles wired on January 2, 1941, that "the Department feels strongly that it would be most undesirable under the present conditions to have a public controversy between the Venezuelan Government and the Gulf Oil Company." Observing further that "the international implications will be obvious to you," Welles urged Corrigan to keep the dispute within "friendly channels." In March, Corrigan reported that the dispute had taken a dangerous turn. Negotiations had broken down. Venezuela insisted on $15 million in back taxes; Gulf offered only $2 million to $3 million. Corrigan scathingly described Gulf's negotiator, James Greer, as "a hard-headed reactionary without much mental flexibility." López Contreras now planned to bring civil and criminal charges against Gulf. The ambassador implored the department to speak with Gulf executives.[27]

The department immediately heeded Corrigan's plea. On April

8, Sumner Welles met with Colonel J. Frank Drake, Gulf's president. While he did not record his conversation with Drake, it seems that Welles informed the Gulf executives that the department would not support the company's position. On the day of the meeting, Philip Bonsal, the new chief of the Division of American Republics, told Welles that "without taking any position as to the merits of the controversy, our feeling that the Venezuelan government had made a painstaking study of the problem and is very far from proceeding in an arbitrary manner should be made clear to Mr. Drake."[28] Drake agreed to compromise, and nine days after his meeting with Welles, Gulf reached an accord with Venezuela that called for a $10 million payment in back taxes, $5 million to be paid immediately in cash.[29] With the settlement, the crisis quickly abated.

In reviewing the dispute, Division Chief Bonsal observed "that the Gulf people have been very fortunate in extricating themselves from a very difficult situation in which their own unreasonableness had placed them."[30] The State Department also had been fortunate that López Contreras willingly dropped charges against Gulf after receiving the payment. Despite the shocks of the Bolivian and Mexican expropriations and its expressed intention in July 1939 to mediate between the Venezuelan government and the oil industry, the department had permitted, once again, a U.S. oil company to jeopardize its foreign policy. Welles had spoken to Drake less than a week before López Contreras planned to make the tax fraud case a crusade in defense of Venezuela's sovereignty. It was, however, the last time during the war that the department would allow the oil issue to drift.

On April 19, 1941, two days after Gulf agreed to pay back taxes, President Eleazar López Contreras ended his term. His regime had been one of transition for Venezuela. While still preserving authoritarian rule, López Contreras curbed the corruption and brutality that characterized the Gómez era. And he demonstrated a modicum of respect for the constitution by limiting his term and adhering to the no-immediate-reelection stricture. López Contreras refused, however, to allow a free national election, and he instructed Congress to approve his handpicked successor. Attempting to give meaning to his pledge to point Venezuela toward democracy, López Contreras permitted his leftist opposition to run a token candidate in the rigged presidential election.[31]

General Isaías Medina Angarita, the minister of war, became Venezuela's new president in April 1941. He was another member of the military clique from Táchira that had ruled Venezuela since 1899. Unlike Gómez and López Contreras, however, the forty-three–year–old Medina had not participated in Castro's capture of

Caracas. A graduate of the Military Academy in Caracas, Medina had spent most of his military career outside his native state. A cultured, intelligent, and humane man, Medina seemed in temperament and outlook to be more of a *caraqueño* than a rugged mountain man from Táchira.[32]

Perhaps swayed by the democratic fervor engendered by World War II, President Medina accelerated the pace of Venezuela's political liberalization. He tolerated critics and allowed opposition parties to organize. He respected freedoms of press and speech. Moreover, he did not imprison or exile a single Venezuelan for political activity. No other Venezuelan administration before Medina's or few since could make that claim. And, perhaps most remarkably, Medina attempted to legitimize his policies with popular support. He traveled throughout the country and frequently spoke at rallies organized to generate public approval of his policies. In effect, Medina's Venezuela had all the trappings of a democratic state, except a basic one—a free national election.[33]

Medina also pressed for economic and social reforms. During his four years in office, his administration introduced changes in education, industry, and finance; modernized labor laws; enacted a social security program; instituted Venezuela's first progressive income tax; and passed an agrarian reform law. As significant as they were for Venezuela, those accomplishments were dwarfed by the reform of the oil laws. The Medina government abolished the Castro and Gómez system and wrote an oil code that governed the foreign oil industry for over thirty years.

Within months after assuming power, President Medina took up the uncompleted task of his predecessor and renewed the campaign for a new relationship with the oil industry. In July 1941, he sent Gustavo Manrique Pacanins, Venezuela's solicitor general, to New York to discuss with the presidents of Gulf and Standard Oil of New Jersey the need for new petroleum legislation. In particular, the Venezuelans wanted increased royalties, conversion of old concessions and contracts to a new law, construction of oil refineries in Venezuela, and technical improvements, such as the conservation of natural gas during drilling. Manrique Pacanins had hoped to deal with officials in New York; he believed that they would be more amenable to change than the local managers. But the presidents pointed out that both sides always operated under the assumption that the local companies were actually Venezuelan concerns and not subsidiaries of foreign companies. Accordingly, negotiations moved to Caracas where Manrique Pacanins and the local managers met and exchanged proposals on new oil legislation. The negotiations continued until November when Manrique Pacanins unexpectedly stopped meeting with the managers.[34]

The Medina government suspended negotiations in late 1941, because it lacked the degree of purpose and direction needed to win concessions from the foreign oil companies. Though friendly toward Manrique Pacanins, the local managers, led by Henry Linam of Standard Oil of New Jersey, continued to insist on their legal rights and the validity of their concessions. The solicitor general argued that change was a political necessity and that superseded any "legal rights." Yet, Venezuelans found it difficult to transform political necessity into reality. They lacked the technical expertise to develop a new set of regulations for a complex industry. The oilmen with their technological sophistication easily countered the Venezuelans' proposals. Divisiveness within the government also handicapped Venezuela. The official normally in charge of petroleum affairs, Minister of Fomento Enrique Aguerrevere, preferred to assert the nation's rights in the courts by continually challenging the oil laws and the validity of Gómez' concessions. President Medina would have to choose between his competing subordinates before there could be a concerted national effort to regulate the petroleum industry. Finally, the government probably softened its attitude toward the industry as petroleum taxes filled the coffers. Unlike 1940, 1941 was a good year for Venezuelan oil. A rapidly mobilizing United States needed Venezuela's oil; its purchases replaced the lost European sales. Production in 1941 was 23 percent higher than the previous year and 11 percent above the 1939 high.[35]

But if Venezuela profited from the vicissitudes of world politics, it also suffered from them. In 1942, Venezuelans learned how fragile a one-industry, export-oriented economy could be. After the United States declared war, Nazi submarines began to prowl Caribbean sea lanes. On February 14, 1942, they torpedoed seven oil tankers off Venezuela's Paraguaná Peninsula. Thereafter, tankers moved only at night in convoy. Less petroleum could be shipped by convoy, and there were fewer tankers available than before the war. The Allies were losing ships to torpedo attacks and deployed their remaining tonnage in the Asian and European theaters of war. With shipping often unavailable, Venezuela's producers had to cap wells. Production in 1942 was 148 million barrels, a 35 percent decline from the previous year.[36]

The effects of the curtailed petroleum business reverberated through the economy. During the calendar year, the government's petroleum revenue declined by 38 percent. Import duties, normally Venezuela's largest source of income and always linked to the level of petroleum activity, fell by 28 percent. Total government revenues amounted to Bs. 287. million or approximately $96 million, 20 percent less than in 1941. In order to cope with the budgetary crisis, President Medina cut wages of government employees, eliminated development programs, drew on treasury reserves, and, when reserves reached dangerously low levels, floated the public debt.[37] He also resolved to raise petroleum taxes.

Apart from the need for new revenues, 1942 seemed a propitious time for Medina to press the foreign oil industry. As they had in the past, developments in Mexico accrued to Venezuela's benefit. On April 17, 1942, Mexico and the United States announced the settlement of the oil expropriation issue. Mexico agreed to pay the U.S. oil companies approximately $24 million plus $5 million interest for their properties. The settlement was a resounding victory for Mexican diplomacy; Mexico had successfully resisted the State Department's position that the expropriation be submitted to third-party arbitration, and it upheld its contention that subsoil rights (oil still in the ground) belonged to the nation and not the oil companies. Mexico paid the U.S. companies approximately one-tenth of what they demanded. During the protracted negotiations, Mexican diplomats had skillfully traded promises of military cooperation for the U.S. acceptance of their plan that a joint commission of experts determine the value of the expropriated properties. Aware of the War Department's belief that air bases in Mexico would be essential in defending the Panama Canal, Secretary Hull and Under Secretary Welles gradually concluded that to prolong the bitter controversy would endanger both the Good Neighbor policy and hemispheric solidarity. They had also grown weary of the oil companies' intransigence. Their acceptance in November 1941 of the joint commission plan signified a triumph not only for Mexico but also for Josephus Daniels and Laurence Duggan in their bureaucratic struggle over the proper U.S. response to the expropriation. For Venezuela, the April announcement of the settlement indicated that there were now limits to the State Department's support of U.S. oil companies.[38]

President Medina initiated the second phase of his oil legislation campaign in mid-1942. In a speech from the presidential palace on July 16, Medina promised to revise the petroleum laws "to assure the state of a greater and more just participation in the wealth of its subsoil."[39] This public pledge underscored the firmness of Medina's resolve. Medina's next moves, even before speaking with the oil companies, were to inform both Ambassador Corrigan and British Ambassador Gainer of his intentions and to send Manrique Pacanins to Washington with a letter for President Roosevelt that explained his country's position.[40] Such a battle plan revealed that Medina interpreted the Mexican settlement to mean that he could expect State Department sympathy and perhaps support in his campaign to write new petroleum legislation. In addition, it demonstrated that he had decided to choose Manrique Pacanins' negotiated settlement methods over Minister of Fomento Aguerrevere's wish to drag the oil companies through the courts. Medina ended the bureaucratic bickering by replacing Aguerrevere with Eugenio Mendoza, a minister who would support Manrique Pacanins.

As he undoubtedly hoped it would, Medina's letter to Roosevelt brought the United States into the impending negotiations. In his letter, Medina reaffirmed Venezuela's solidarity with the United States and disclaimed any intention of nationalizing the oil industry, although he made an oblique reference to expropriation by observing that "the present situation, unfavorable for the country already, could some day become intolerable, bringing with it serious injury to the mutual interests of the nation and the oil companies." In his September 14 reply, drafted by Assistant Secretary of State Adolf A. Berle, Roosevelt acknowledged Venezuela's support, noted approvingly that Medina intended "to deal fairly" with the oil companies, and expressed confidence that a satisfactory settlement would soon be reached. Under Secretary Welles backed Roosevelt's polite, encouraging letter by informing Manrique Pacanins that "the Department of State would do whatever it appropriately could to facilitate a friendly adjustment."[41]

Medina followed his entreaties to the State Department with a stern warning to the oil companies. On August 26, he summoned the local managers and informed them that proposals submitted to Manrique Pacanins in 1941 fell far short of the government's expectations. Since conferences had not produced results, the president warned that he would pursue "different and more direct methods."[42] Implicit in his warning was the threat that he would legally challenge oil concessions obtained during the Gómez era. It is uncertain whether Medina actually contemplated legal redress or was merely collecting bargaining chips for future negotiations. In any case, his warning stirred the State Department.

The department labored on the oil problem on three fronts. In Caracas, Ambassador Corrigan met with the foreign minister on September 5 and successfully urged him to postpone any unilateral action against the companies. At the same time, the ambassador belittled the oil industry's behavior in a despatch to Washington. He noted that local officials were forlornly hoping that the days of Gómez would return when "a policy more attuned to sympathy with a nation striving to regain its birthright of democracy would serve a better purpose." The department responded to Corrigan's despatch by meeting with officials of Standard Oil of New Jersey on September 23. Welles and Bonsal stressed the vital need for friendly negotiations and implored Standard officials to recall Henry Linam, whose long association with Gómez had made him anathema to many Venezuelans.[43]

While Corrigan, Welles, and Bonsal urged calm on both sides, Max W. Thornburg began to arrange a settlement. Thornburg, the department's petroleum advisor since 1941, was a former employee of Standard Oil of California. A refining engineer by training, he had gained administrative and international experience as a director of a consortium of U.S. oil companies exploring in the Bahrein Islands in the Persian Gulf. With Welles' permission, Thornburg had informal meetings in his apartment with both Manrique Pacanins and oil executives in an attempt to draft provisions for a new oil law. Like other members of the department, Thornburg had changed his position on oil problems. He had opposed the Mexican joint commission plan because it would create an "unfortunate precedent" and might jeopardize U.S. holdings in other countries. But he now favored conciliation, for, as he told Welles and Bonsal, "our primary interest is that Venezuelan oil remains available for the war."[44]

With serious negotiations underway, Manrique Pacanins returned to Caracas in late October and reported to President Medina. Through the United States Embassy in Caracas, he regularly received reports from Thornburg on the progress of meetings with oil executives. In effect, Thornburg was writing Venezuela's new oil law. His entrance into the controversy removed the crisis atmosphere from industry-government relations. Both sides now expected a law to be written along fair principles and by fair methods.

In late November, a further relaxation of tensions occurred after Standard Oil recalled Henry Linam. Linam had stubbornly upheld the validity of Standard's concessions and refused to consider any basic changes in the 1922 oil law. As British Ambassador Gainer described him, Linam was a self-made, uneducated man who detested Roosevelt and the New Deal, "and would, if he could, suppress all individual liberties 'among the masses,' reserving them solely to those who, like himself, have shown their ability to make and maintain a position for themselves."[45] During his nearly two decades in Venezuela, Linam had secured most of Standard's concessions, many allegedly with bribes. His anachronistic presence in post-Gómez Venezuela had been a continuing source of irritation to many. As one official had complained, it was as if a member of Gómez' cabinet remained in power.[46]

Linam's removal coincided with a shake-up in Standard Oil's hierarchy. Just after recalling Linam, President William S. Farish, who had fiercely fought the Mexican expropriation, died of a heart attack. Ralph W. Gallagher became the new president and his vice-president was Wallace Pratt, who had been meeting with Thornburg and Manrique Pacanins.[47] Pratt's appointment virtually guaranteed a peaceful, negotiated settlement of the Venezuelan oil problem. Standard's acceptance of an agreement was essential, for it was the recognized leader of the industry; it produced nearly half of Venezuela's petroleum, and it was the largest holder of concessions. Gulf, which produced slightly more than 10 percent of Venezuela's petroleum, was expected to follow Standard's lead.

Standard's decision to accept changes in its concession rights also meant that Royal Dutch–Shell would have to compromise. During the 1930s, Standard and Shell, who together produced almost 90 percent of Venezuela's petroleum, had become friendly business competitors. Both corporations found that Venezuela held rich petroleum reserves, and they gradually tempered their intense rivalry of the early 1920s. Moreover, in the face of anti-industry sentiment that followed Gómez' death, both saw it was in their mutual interest to band together in a "united front" to counter political and legal challenges to their concessions. But in late 1942, Shell officials learned that, for the second time, Standard Oil, under diplomatic pressure and perhaps in its own interest, had broken industry solidarity.

Shell and Standard had worked together against Mexico after the 1938 expropriation. They had consistently denied the legality of expropriation, claimed subsoil rights as their own, and organized a worldwide boycott of Mexican oil. Accordingly, Shell officials were distressed to learn in early September 1942 that Standard planned to accept the expropriation and forfeit its claim to subsoil rights by accepting the Mexican-U.S. settlement of $24 million. President Farish apologized but noted "that he was of the opinion that if his company showed any intransigent attitude over Mexico, they would certainly have no chance of getting any help whatever from the government with Venezuela." As Britain's Foreign Office and Shell quickly surmised, Standard had made a tactical sacrifice.[48] Compared to Shell's holdings, Standard had a minor stake in Mexico. Prior to the expropriation, it produced less than one-third of Mexico's oil while Shell produced over 60 percent. Their relative positions in Venezuela were almost exactly the opposite. Moreover,

Standard had extensive holdings in the proven and largely untapped fields of eastern Venezuela. Production in Mexico's fields, on the other hand, had been declining since 1921 (see Appendix, Table C). It was to Standard's advantage to concede its loss in Mexico and hope for diplomatic aid in Venezuela.

A month after Farish apologized, Shell and the Foreign Office received a second shock from the U.S. oilmen. On October 5, 1942, Welles summoned Lord (Edward) Halifax, the British ambassador to the United States. The under secretary lectured Halifax on the history of the Mexican expropriation. "Most difficulties in Mexico," in Welles' opinion, "had been largely brought on by the unwisdom of the British and American Governments, who tended to treat Mexico with thinly concealed indifference." He feared that "the same influences were beginning to operate in Venezuela." Venezuela disclaimed any desire to nationalize the industry, but it might take "legal action to get rid of certain clauses in the oil concessions which were so unreasonable as to have been secured by something like sharp practice or corruption." Welles warned Halifax that "the sands were running out" and "that unless the companies acted both quickly and wisely, recognising what he felt to be a claim of the Venezuelan Government founded in justice and equity, they would run into bad trouble."[49]

Halifax suffered Welles' diatribe in silence. Both he and the Foreign Office knew what Welles meant. After the Mexican expropriation, the Foreign Office had taken a harsher stand than had the State Department. It had challenged Mexico's right to expropriate, and it had demanded assurances that Mexico would quickly and justly compensate Shell. Mexico regarded these tactics as interference in its domestic affairs and broke relations with Great Britain in May 1938. The oil issue remained unsettled more than four years after the expropriation.[50] In summoning Halifax, Welles wanted to inform the British that the State Department had already decided on the merits of Venezuela's case and that the Foreign Office must not use heavy-handed diplomatic pressure against Venezuela as it had against Mexico.

In a slightly more subtle manner than in its approach to the Foreign Office, the department informed Shell that it would have to accept a U.S. solution to the Venezuelan problem. In late October, Max Thornburg called on Frank J. Hopwood, Shell's representative in the United States. He mentioned to Hopwood that the U.S. oil companies were submitting their proposals for a new oil law through the State Department; he invited Shell to use the department's facilities. Thornburg later asked Hopwood to meet Manrique Pacanins in his apartment. As Hopwood related to his superiors in London, Manrique Pacanins seemed to talk about the proposed legislation with an air of confidence and assurance.[51] The Venezuelan's cockiness came, of course, from the knowledge that he spoke with the approval of the United States.

The Foreign Office and Shell gloomily concluded that only the United States could "carry any real and possibly decisive weight" with Venezuela and agreed to follow Thornburg's lead. They acknowledged their painful but unavoidable humiliation at the hands of the State Department. Since the end of World War I, the British had been acutely aware that they were losing their historic economic and political preeminence in South America to North Americans. Welles' stinging rebukes and Thornburg's adroit maneuverings served to remind the British of that decline.[52] After World War II, the British would routinely defer to the judgment of the State Department and Standard of New Jersey on oil matters in Venezuela.

By excluding the British, Welles and Thornburg had simplified and perhaps accelerated the pace of negotiations. But their actions were probably unnecessary. The Foreign Office had no desire to jeopardize Britain's major source of petroleum in the middle of a war. Moreover, Ambassador Donald Gainer agreed with his good friend and colleague Frank Corrigan that the oil companies would have to recognize their social responsibility and contribute more of their income to the government. Shell officials also had considered taking a conciliatory attitude toward a new law. Indeed, Th. W. van Hasselt, the new manager of Shell's holdings, had been willing to bargain about a new law when Manrique Pacanins first approached the companies in 1941. In the interests of the "united front," van Hasselt had, however, adopted Henry Linam's strategy and insisted on Shell's acquired rights during the initial negotiations.[53]

With the Anglo-American diplomatic wrangling over, final negotiations for a new oil law began in Caracas on December 1, 1942. Standard and Gulf sent representatives to bargain directly with the government. The key oilman was Arthur Proudfit, the new manager of Standard's Venezuelan subsidiaries. During the meetings, Venezuela relied on the technical advice of two U.S. citizens, Arthur Curtice and Herbert Hoover, Jr. While in the United States, Manrique Pacanins had hired these geological engineers to ensure that Venezuela could match the companies in any technical discussions about the oil industry. Also present in Caracas during December was Max Thornburg. President Medina had invited Thornburg to Caracas, and, with Welles' permission, Thornburg accepted with the understanding that he would be an unofficial observer. The department wanted to appear aloof from Venezuela's domestic concerns. Thornburg's "unofficial" status was polite, diplomatic nonsense. While in Caracas he remained on the State Department's payroll. And government and industry worked on writing a law from an outline that he had previously presented to both sides.[54]

The negotiations proceeded peacefully and successfully for the next two months with only one disturbing problem. President Medina had insisted that Venezuelan crude oil be refined on Venezuelan soil. The foreign oil companies refined about 95 percent of Venezuela's production outside the country. Shell and Standard transported their crude to their refineries on the Dutch West Indies islands of Curaçao and Aruba, while Gulf shipped its production directly to the United States. Venezuela had only a few small refineries that Standard and Shell operated for their own and Venezuela's consumption. This was another unhappy legacy from the Gómez era. Shell and Standard of Indiana (purchased later by Standard of New Jersey) had located their refineries in the Dutch West Indies in the late 1920s for essentially political reasons. Anticipating violence after Gómez' death, they had decided that it would be safer to construct the costly refineries in the politically stable Dutch colonies.[55] Gómez had allowed the companies to build the refineries away from the oilfields because he feared that if the oil-producing western states, such as Zulia, became too prosperous they might declare indepenence from Caracas.[56]

Whenever Venezuelans listed their grievances against the oil industry, they always included the refinery issue. They disliked having European colonies so near their borders, and it irritated them to know that their oil contributed to Dutch West Indian prosperity. Curaçao's population had doubled and Aruba's had tripled since the refineries had been built. Venezuelans argued that the new jobs which had been created by the refineries belonged to them.[57]

Expressing those national feelings on November 16, two weeks before the beginning of negotiations in Caracas, President Medina publicly announced his wish to move the Aruba and Curaçao refineries to Venezuela. As Ambassador Corrigan later remarked, Medina's announcement was a "bombshell." Neither the State Department nor the oil companies had anticipated the demand, and both immediately rejected it. The companies refused to consider packing up such valuable properties. And the department would not countenance any disruption in the flow of petroleum, crude or refined, to the war effort. Medina quickly realized that he had made a tactical error in committing himself before the nation; a way had to be found to save face. After some haggling, Shell and Standard agreed to build in Venezuela a refinery with a daily capacity of forty thousand barrels of refined products. The companies would not be obliged to begin construction until after the war, however, since steel was in short supply.[58]

With the last hurdle cleared, negotiators quickly completed their work. In late February 1943, President Medina submitted the lengthy and complicated legislative proposals to Congress, and on March 13, the Congress enacted them into law. The oil law of 1943 reformed the industry. Its basic provisions called for the conversion of old concessions to the new law, an increase in royalties from the previous rates varying from 7.5 to 11 percent to a new minimum rate of 16.66 percent, the boosting of exploration and exploitation taxes and the fixing of those taxes on a progressive scale designed to encourage production, the elimination of the industry's right to import duty-free essential materials, and the promotion of domestic refining. The major intention of the new law was to increase the nation's oil income. Both sides expected that with the new royalties and taxes Venezuela would receive one-half of the oil industry's profits and that its income would increase by 80 percent.[59]

The new oil legislation was indeed a financial boon for Venezuela. Compared to 1941, a prosperous year, total government income in 1944 was almost two-thirds higher, even though oil production had only increased by 11 percent. Thereafter, with the foreign oil industry's continuing, rapid expansion of production, the government's income grew at a staggering pace. In 1947, for example, total income, allowing for inflation, was 358 percent higher than in 1941.[60] Venezuela became one of the few Latin American countries that could underwrite social and economic development programs.

The new law also had significant ramifications outside Venezuela. The "50-50" split of profits became standard in other oil-producing nations. After World War II, Persian Gulf nations negotiated contracts with the U.S. and British oil companies similar to the Venezuelan law. Indeed, some of the key figures in the Venezuelan negotiations—Thornburg, Curtice, and Hoover—helped write Iran's oil code.

Aside from its tangible effects, the oil code of 1943 is noteworthy, for it has been interpreted as a watershed in U.S. diplomatic history. Historian Bryce Wood has argued in his influential study of the Good Neighbor policy that the State Department's successful intercession in the government-industry dispute "established as a fundamental principle of the Good Neighbor policy that there was a national interest of the United States in its relations with Latin

America, different from and superior to the private interests of any sector of business enterprise or of business enterprise as a whole." Before 1939, "the main business of United States diplomats in Latin America . . . was the defense of the economic interests of their co-citizens." But after July 1939, when Welles accepted Duggan's arguments, the State Department decided that the security of the United States far outweighed the interests of the oil companies. Moreover, by participating in a process that brought substantial economic benefits to Venezuela, the State Department "came close" to recognizing that it had "a measure of responsibility for the welfare of Venezuelans."[61]

As Wood has admitted, it "may seem both simple and banal" to make much of a policy that gave priority to wartime oil needs over the acquired rights of the oil companies. But he has argued that the decision to formulate a new policy represented a break from the department's vigorous defense of the oil companies' holdings in Bolivia and Mexico.[62] Wood can be questioned for not explaining why, despite its new oil policy, the State Department allowed the oil issue to drift until April 1941 and for not considering that the new policy arose as much from the department's abject failures in Bolivia and Mexico as from a desire to promote social progress in Venezuela. In any case, the core of Wood's thesis transcends questions of timing or origin. His essential argument is that the State Department sacrificed the interests of U.S. businessmen in order to attain foreign policy goals. In order to determine whether U.S. foreign policy and the oil industry's interests were in conflict, as Wood implies, and to gain insight into the nature of the Good Neighbor policy, it is necessary to examine the aspirations and motives of those involved with the 1943 oil law: the Medina government, the anti-Medina political factions, the U.S. oil companies, the Royal Dutch–Shell Oil Company, the Foreign Office, and the State Department.

In seeking a new oil code, President Medina had the obvious leverage of the United Nations' wartime need for oil. Medina astutely played on the State Department's fears in bringing it into the negotiations. But the president also had handicaps. In mid-1942, Venezuela was in deep economic trouble. Medina desperately needed a new agreement with the foreign oil companies, but he lacked the ultimate stick to prod the companies into accepting the new legislation. Both he and the companies knew that Venezuela could not afford expropriation. In all sectors of the oil business—prospecting, production, refinement, marketing, and distribution—Venezuela did not have the trained people or expertise to replace foreigners. The government's need to hire the U.S. experts, Curtice and Hoover, revealed Venezuela's weaknesses. Expropriation would have turned a troubled economy into a bankrupt one.[63]

In lieu of the stick, Medina offered the foreign oil industry some bright carrots to lure them toward an agreement. In return for its acceptance of the new law, the industry received new forty-year leases on its holdings; the right to acquire new concessions with forty-year leases; Venezuela's promise to drop pending or proposed lawsuits against the industry for past practices; a unified, workable, technically sound oil code; and the government's pledge to launch a public relations campaign in behalf of the industry. The oil industry particularly wanted the first two of Medina's offers. During the late 1930s, geologists had gradually concluded that Venezuela would be a significant producer for decades. Even though some Castro- and Gómez-era concessions would not lapse until the 1950s and 1960s, the companies would not have time to tap all the oil riches. Moreover, geologists had found vast new deposits in unleased lands in eastern Venezuela. These promising new fields held a type of high-gravity, light crude that could be refined into gasoline. In contrast, the Maracaibo oilfields produced a low-gravity, heavy crude mainly suitable for products like home-heating oil, which sold at a lower price than gasoline did. Accordingly, the oil companies took advantage of the new code to renew for forty years six million hectares of their old concessions and to lease in 1944 nearly six million hectares of new concessions.[64]

The 1943 oil law was as much a boon to the foreign oil industry as to the Venezuelan treasury. As geologists had predicted, Venezuela expanded its oil production nearly every year after 1943 with a peak production of 1.35 billion barrels in 1970. During that period, Venezuela retained its rank as the world's leading oil exporter. Such output meant substantial profits for the oil industry. Between 1945 and 1970, after-tax profits amounted to over $11 billion. The companies' annual profits normally averaged between 20 and 25 percent of the value of their net fixed assets.[65]

Despite Venezuela's alluring prospects, the oil companies, particularly Standard Oil of New Jersey, did not immediately seize the opportunity to bury the alleged misdeeds of their past and establish their businesses in a new stable and prosperous relationship with the government. The State Department had to pressure Standard into reconsidering its position in Venezuela. Even after the Mexican debacle, some Standard officials had confidently expected diplomatic assistance. In July 1941, Standard's Thomas R. Armstrong pointedly reminded Livingston Satterthwaite of the Division of American Republics "that the uninterrupted production of oil was vital to the defense of the British and this hemisphere . . . and that the Department might wish to take the defense aspect into consideration in using its good offices to facilitate an equitable solution." A month later, Satterthwaite reported that he had heard from "reliable sources" in Trinidad that the oil companies expected the aid of U.S. battleships if a crisis developed.[66] The department's sympathetic reception to Medina's and Manrique Pacanins' overtures firmly quashed, of course, such hopes.

Yet, in refusing to exert pressure on Venezuela, the State Department had not necessarily moved against Standard Oil. Instead, its policy strengthened the position of Standard executives, such as C. H. Lieb, Wallace Pratt, and Arthur Proudfit, who argued for new approaches toward the host country. They pointed to both the lessons of the Mexican experience and Venezuela's oil potential in persuading their colleagues to negotiate with Venezuela. In their campaign they were aided by an influential stockholder, Nelson Rockefeller. Prior to his appointment in August 1940 to head the department's wartime informational agency, the Office of the Coordinator of Inter-American Affairs, Rockefeller had toured the oilcamps in Venezuela, warning that if the company was not socially responsible "they will take away our ownership." In altering their company's stance, the oil executives and Rockefeller triumphed over old Venezuelan hands like Linam.[67]

The Department of State also presented a powerful bargaining chip to those oil executives who favored conciliatory policies. Diplomats informally assured them that, if the companies accepted a new law, they could expand production, for the United States would try to increase allocations of scarce oilfield equipment to Venezuela as well as provide more naval protection for tankers. Indeed, oil production in 1945 was 323 million barrels, as compared to 148 million barrels in 1942. President Roosevelt also helped by requesting in 1944 that the Office of War Mobilization find some steel for Venezuela so that the companies could fulfill their promise to build a refinery.[68]

In both pressuring and inducing Standard to negotiate, the State Department knew that the company was not left in the hands of a radically nationalistic government. The department recognized at the outset of negotiations that the Medina administration was "not hostile to the companies," and Medina and Manrique Pacanins continually assured the department that their government would treat the companies fairly. As had López Contreras, President Medina simply wanted a larger share of the companies' profits. During the secret negotiations in Caracas, he and his officials consulted with the industry on proposed provisions of the code; the companies later admitted that the negotiations had been fair and friendly.[69] Medina broke his pledge to work with the industry only once when he called for the moving of the Aruba and Curaçao refineries to Venezuela. He quickly regretted his blunder, and he balanced it with public praise for the companies' cooperative spirit.

The reaction of dissident Venezuelans to the 1943 law underscored Medina's moderation. Medina allowed opposition parties to organize, and many of the leftist critics, whom López Contreras had suppressed, merged in the new Acción Democrática party, whose members were popularly known as the *adecos*. A major plank of the new party's platform was the demand for a new oil law. But the law that Medina presented Congress displeased the *adecos*. In part, their criticism smacked of jealousy; Medina had accomplished what they thought only the political left could. They jeered that Manrique Pacanins had formerly served as an attorney for foreign oil companies, and they argued that the proposed refinery should have a larger capacity than forty thousand barrels. But beyond petty complaints, the *adecos*, led by Deputy Juan Pablo Pérez Alfonzo, pointed to some disquieting passages and glaring omissions in the new law. Pérez Alfonzo denounced the government's absolution of the companies' past misdeeds, the so-called original sins. In effect, the 1943 law condoned the corruption and tax evasion of the past. But what particularly worried the critics of the law was that it made no provision for Venezuela to enter the oil business. The proposed 1938 law gave the government the right to undertake both production and refinement. The 1943 law left the entire oil business and, in effect, all of Venezuela's major economic decisions in the hands of foreigners. The *adecos* shared the government's belief that Venezuela was unprepared to take control of the complex oil business, but they argued that it should gain experience.[70] These warnings and objections achieved little, for Medina allowed only token debate and no significant modifications in the law. He had won for Venezuela substantial economic benefits while reaffirming the foreign oil industry's control over his nation's economy.

The law of 1943 affected not only Venezuelans and U.S. oilmen but also the British. It could be argued that the State Department and, particularly, Welles prevented Britain from exerting diplomatic pressure on Medina to moderate his demands. British diplomats loathed Max Thornburg as a "new dealer and advocate of the good neighbour policy" who would "consider no price too high to obtain a long-term peace." In fact, the U.S. companies and Shell believed that Thornburg allowed the Venezuelans to demand everything the companies could possibly afford. But the Foreign Office and the Royal Dutch–Shell did not object so much to the ends which Thornburg sought but rather to his tactics. Th. W. van Hasselt, the manager of Shell's Venezuelan holdings, thought it more fruitful to negotiate with the government in piecemeal fashion, instead of seeking, as Thornburg wanted, a long-term, overall settlement. But though he initially disagreed with Thornburg's tactics, he approved of reducing tensions. As early as 1940, van Hasselt wanted to accommodate the government. He believed it essential to strengthen such moderates as López Contreras and Medina to help them ward off leftist extremists. In the end, van Hasselt, Shell, and the Foreign Office expressed satisfaction, albeit grudgingly, with Thornburg's labors.[71] He had restored harmony and preserved Shell's place in Venezuela.

Against the backdrop of the aspirations and concerns of the Medina government, the Acción Democrática party, and the U.S. and British-Dutch oil companies, the State Department's Venezuelan policy can be analyzed. The department refused to support contracts that the oil companies had obtained prior to 1935. As Under Secretary Welles wrote to President Roosevelt, "the arrangements under which the oil industry got its start in Venezuela and the conditions then prevailing are no longer applicable to an established industry operating one of the richest and best proven oil fields in the world."[72] As such, the department modified certain interests of U.S. and British businesses in order to placate a valuable wartime ally, to ensure that Venezuelans received an equitable share of their own bounty, and because it now considered those contracts of dubious legal and moral merit. But the settlement that took away the old rights of the oil companies also conferred substantial new benefits upon them. Moreover, influential oil executives had been prepared to sacrifice those old rights in order to end the nettlesome bickering and to gain access to promising unleased land. The State Department had not imposed a settlement on the oil companies but rather facilitated negotiations between a moderate Venezuelan government and a compromise-minded oil industry. As one scholar has recently concluded in his analysis of the Good Neighbor policy, "the State Department never abandoned its defense of private property."[73] Indeed, in 1945, a group of eleven U.S. oil companies would cite the 1943 law as "an outstanding example of the favorable economic results to both a country and private business of an 'open door' policy backed up by equitable and stable laws."[74]

Bryce Wood's use of the oil law of 1943 to define the State Department's attitude toward U.S.-owned property abroad and to demonstrate the maturation of the Good Neighbor policy needs revision, for U.S. foreign policy and business interests did not conflict during World War II. The State Department realized that it had to satisfy both the Medina government and the oil industry. As Welles explained to President Roosevelt, the United States had to prevent "a state of public opinion in Venezuela contrary to foreign interests," for it would threaten the continued production of oil. Yet, the interests of the oil companies could not be readily discarded. Only the established oil companies could produce Venezuela's oil. The oil companies had "to continue operating on a satisfactory basis"; nothing could be allowed to impair their "efficient functioning."[75] The department-sponsored oil law helped to fulfill the needs of both sides and thereby kept the oil flowing.

The State Department had accomplished all its foreign policy goals for Venezuela during World War II. With the judicious use of military aid, loans, credits, and trade advantages, it blunted any Axis designs on Venezuela and won the Caribbean nation's political and military support. And, with its new oil policy, it kept Venezuela's oil available for the Allies. The department had not, however, achieved a lasting peace with Venezuela. The new oil code had reinforced Venezuela's dependence on the United States, the foreign oil companies, and a capricious world economy.

5. The *Trienio*, 1945–1948

Between 1945 and 1948, the United States encountered a spirited Venezuelan government that questioned its foreign economic policies. Idealistic civilian politicians who wanted a modern, democratic, independent Venezuela governed the nation during the immediate postwar years. These leaders promoted national development projects designed to free the nation from its traditional dependence on the industrial world for capital and manufactures. They also began to speak of the day when Venezuela would operate its oil industry. The United States welcomed the birth of political democracy in Venezuela but argued strenuously against economic nationalism. In the view of the United States, further infusions of U.S. capital and technology would assist Venezuela in attaining the peaceful, prosperous society that its new leaders so earnestly desired.

On October 18, 1945, the crash of aerial bombs, mortars, and grenades reverberated through Caracas. A coalition of disaffected junior military officers and ambitious civilian politicians attacked the government of President Isaías Medina Angarita. Following well-conceived plans, the insurgents quickly crushed the presidential guard, seized the presidential palace, and captured the president. By October 22, they had subdued both the capital and the interior. A new government, pledged to establish democracy in Venezuela, would guide the country for the next three years.

The October 18 rebellion was a watershed in Venezuelan history. It led to Venezuela's first popularly elected presidential administration, and it brought to national attention men who would dominate politics in the 1960s and 1970s. In addition, it put in power leaders determined to control the nation's burgeoning foreign oil industry. But the rebellion also had the effect of directly reintroducing the armed forces into politics, after a ten-year hiatus since the death of Gómez in 1935. Military officers who participated in the October

18 rebellion later led the *golpe de estado* that ousted President Rómulo Gallegos on November 24, 1948. Thereafter, Venezuela suffered a decade of military rule that rivaled the Gómez regime for brutality and corruption.

The catalyst for the October 18 rebellion was the crisis over the choosing of a successor to President Medina. The constitution barred a president from immediate reelection; a new president was scheduled to take office in April 1946. Medina had permitted political activity, and the hope grew among Venezuelans, particularly among members of the opposition Acción Democrática party (AD), the *adecos,* that Medina would allow a democratic national election. While not disposed to relinquish the power to choose his successor, Medina proposed a solution in mid-summer 1945. Medina's party, Partido Democrática Venezolano, nominated the respected ambassador to the United States, Diógenes Escalante. The presidency of another native of Táchira would seemingly ensure Medina's continued influence in the government and provide a convenient springboard for Medina's return to office in 1951. Yet, Escalante might placate the *adecos,* since he would be the first Venezuelan president in over a half century who did not have a military background, and he had pledged in an interview with AD officials, Rómulo Betancourt and Raúl Leoni, to press for constitutional changes guaranteeing direct national elections and universal suffrage. Moreover, Escalante's nomination, supported by Medina and the *adecos,* would discourage the former president, Eleazar López Contreras, from considering a return to power. López Contreras had broken with Medina, his handpicked successor, over the liberal policies of the past four years. The *adecos,* many of whom had been exiled by López Contreras in 1937, feared that a resurrected López Contreras regime would retard Venezuela's progress toward complete democracy. The Medina-AD coalition collapsed in August 1945, however, for shortly after his return from Washington, Ambassador Escalante suffered a nervous collapse.[1]

Escalante's withdrawal from the campaign confused the question of presidential succession. After deliberating, Medina announced on September 11 his choice for president—Minister of Agriculture Angel Biaggini, who was both politically inexperienced and relatively unknown. The *adecos* interpreted Medina's move both as a renunciation of their earlier agreement and as a thinly disguised attempt to preserve power through the undynamic Biaggini. After his choice was rejected by the *adecos,* Medina looked for support in a way that only increased tensions. In October, he legally recognized Venezuela's outlawed communist party, Partido Comunista de Venezuela (PCV), in return for its backing of Biaggini. Medina's recognition of the PCV outraged Venezuelan conservatives, particularly López Contreras and his followers. Ominous rumors circulated that the ex-general might return to power through armed revolt.[2]

López Contreras was not, however, one of the October 18 conspirators. In the summer of 1945, junior military officers, who were members of the secret military lodge Unión Patriótica Militar, plotted the overthrow of Medina. The junior officers' anger derived largely from professional concerns. They protested their meager pay and limited opportunities for promotion and decried Venezuela's low military standards and training. Many of the young officers had studied abroad, particularly in Peru and the United States, and they yearned for the respect accorded the military in other nations. Their political aspirations coincided with their professional objectives. They favored a democratic government, but, in particular, a government staffed by Venezuelans of ability, honor, and integrity. The junior officers charged that the old guard's incompetent, dishonest leadership had been the ruin of both the government and the military.[3]

The *adecos* had been aware throughout the summer of 1945 of the junior officers' machinations. With Escalante's withdrawal and Medina's promotion of Biaggini, many *adecos* despaired of "an evolutionary solution to the complex political situation in the country." On the night of October 16, AD leaders enlisted in the conspiracy; they concluded in part that the officers would probably rebel even without civilian support. The October 16 decision fatefully violated, however, two of the party's cardinal principles—opposition to unconstitutional changes of government and removal of military influence from political life.[4]

The conspirators had tentatively scheduled the insurrection for December, but a partial disclosure of the plot forced them to strike in October. The young officers, approximately three hundred strong, carried out the *golpe*, with the *adecos* limited to disseminating propaganda. A seven-man junta, composed of four *adecos*, two officers, and an independent civilian politician, claimed governmental authority on October 19. Rómulo Betancourt, a hero of the "Generation of '28" and a founder of AD, headed the junta. The military insurgents ceded to their civilian counterparts control over all governmental matters except those related to defense and national security, for they wanted "to dissipate any thought that they had instigated the coup out of militaristic ambitions." In its initial statements, the AD-dominated junta justified the revolution by promising a new constitution and free national elections.[5]

While the Department of State was unaware of the conspiracy, it had been astutely preparing for political change in Venezuela. During the Medina years, the embassy in Caracas informed Washington of the political fortunes of Acción Democrática and, especially, Betancourt. Moreover, both the department and the embassy courted AD leaders. When, for example, Betancourt and Raúl Leoni journeyed to Washington to confer with Ambassador Escalante in July 1945, department officials, upon the embassy's advice, had an informal meeting with the two men. The department knew that the *adecos* had severely criticized the oil law of 1943, and it wanted to learn if an AD-dominated government would respect the law. Betancourt and Leoni assured the diplomats that the economic and social progress they envisioned for Venezuela depended upon revenues that only the foreign oil industry could produce. As they had with embassy officials, Betancourt and Leoni impressed Washington officials with both their reasonableness and their dedication to a democratic and socially just Venezuela.[6]

After subduing the country, the junta promptly asked for diplomatic recognition from the United States. The junta undoubtedly judged the recognition of the United States vital to its survival in view of both the U.S. preponderant political and military influence in the Western Hemisphere and the economic stake its corporations held in Venezuela. On October 23, Betancourt met with Standard Oil of New Jersey, Gulf, and Royal Dutch–Shell executives and assured them of the junta's intention to honor the oil code. The oil companies were asked only to concede one thing—a readiness to recognize and bargain with the oilworkers' union. Betancourt reiterated that promise in a statement to the industry publication, the *Oil and Gas Journal*. With those pledges coupled with Betancourt's further assurances in conversations with Ambassador Frank P. Corrigan, the department decided to recognize the junta. It rejected the requests of both the British Foreign Office and the oil companies to tie recognition to a formal pledge to uphold the 1943 law. Indeed, the department seemed eager to recognize a government that promised to respect U.S. investments, support the United Nations, and establish "real democracy in Venezuela, as in the United States." On October 30, 1945, the United States recognized the junta.[7]

After winning recognition by the United States, the AD-dominated junta began to transform Venezuelan society. In rapid succession, the junta decreed reforms that expanded the public sector's role in education, health and sanitation, agricultural modernization, land redistribution, and economic diversification. And it fulfilled its initial promise to establish democracy in Venezuela. A year after the rebellion, Venezuelans elected delegates to write a new constitution, and in October 1947, they exercised their new right of universal suffrage and freely elected novelist and *adeco* Rómulo Gallegos as president in a multicandidate election. As seemed almost predictable from the course of diplomatic recognition, this three-year experiment in political democracy, economic nationalism, and social justice, the *trienio*, affected both the diplomatic ideals and the economic interests of the United States.

In a variety of ways, the United States aided the junta in establishing and preserving democratic reforms. The department and the embassy were helpful, for example, in defusing counterrevolutions. Ambassador Corrigan, who had served in Caracas since 1939 and had contacts with most political factions in Venezuela, frequently mediated between the junta and its opposition. In particular, he sent Allen Dawson, the counselor of the embassy, to Colombia in late 1946 to meet with the exiled López Contreras and to gain for the junta the ex-president's promise not to participate in or lend his name to any conspiracy. Department officials similarly aided the junta by monitoring the movement of arms in the Caribbean area. It instructed officials in Florida to arrest individuals allegedly selling arms to the Dominican Republic. The junta continually feared that rightist Venezuelan factions would use the Dominican Republic, with the blessing of dictator Rafael Trujillo, as a base for an armed attack on the nation.[8]

Besides attempting to prevent a civil war, the department strengthened the junta by providing military aid. At the close of World War II, the State, War, and Navy departments designed an arms standardization program for Latin America. The United States promised to provide Latin American nations with arms and military missions at bargain prices if they adopted, as standard, U.S. equipment and military doctrine to the total exclusion of non-American sources. The departments expected the program to expand U.S. political and military influence in the Western Hemisphere and perhaps help keep the U.S. arms industry vigorous.[9] For Venezuela, the program had immediate political significance. State Department officials frequently debated the wisdom of providing war matériel to a country in the midst of a political revolution. But they generally concluded that limited amounts of U.S. military aid would enhance the junta's prestige, for, as Counselor Dawson noted, "in a country as sensitive to relations with the United States as is Venezuela" military aid would serve "as an indication of direct support for the Revolutionary Junta." Moreover, some department officials and AD leaders believed that military aid would strengthen civilian rule in

Venezuela by keeping young military officers busy pursuing their professional ambitions. Accordingly, the United States established in 1946 air and ground military missions in Venezuela and supplied tactical and transport airplanes and equipment for one infantry battalion and one light field artillery battalion.[10]

A set of interrelated policies and individual goals of U.S. diplomats explains the Department of State's willingness to help the junta. Throughout the twentieth century, such as at inter-American conferences, Washington perfunctorily proclaimed its support for democracy in Latin America. This commitment to the right of peoples to choose freely their government was also expressed, for examples, in the Atlantic Charter (1941) and the United Nations Charter (1945). The United States maintained, of course, warm relations with Latin American despots, such as Trujillo and the Somoza family of Nicaragua, who aped U.S. foreign policy and welcomed investments. Nonetheless, the public statements of AD encouraged the United States to pursue its professed foreign policy goal of a democratic Latin America. Both before and while they were in power, the *adecos* openly expressed admiration for the U.S. electoral process, the social policies of the New Deal, and the U.S. crusade against the Axis powers. AD leaders reiterated those views in private conversations with U.S. officials.[11] In effect, the *adecos* seemingly wanted to convert Venezuela to what the administration of Harry S Truman perceived the United States to be—a democracy committed to the social welfare of its citizens.

Beyond appreciating the *adecos'* apparent desire to emulate the United States, key State Department officials favored aiding AD for personal, political, and professional reasons. Both during and after World War II, the State Department, especially its Latin American section, fractured into competing, often hostile factions. Disputes over Argentina created dissension. Some officials, such as Secretary of State Cordell Hull, believed that Argentina's wartime neutrality was a mask for profascist sympathies. Others, led by Under Secretary of State Welles, argued that Argentina had a long tradition of neutrality and that, in any case, it willingly sold beef and wheat to the Allies. The dispute over Argentina contributed both to an open break between Hull and Welles and to the eventual ouster of the latter from the State Department in 1943. After Welles resigned, Hull conducted a noisy propaganda campaign against Argentina and pressured other states of Latin America to withdraw their representatives from Argentina.[12]

By early 1945, however, the department had shifted its Argentine policy. Hull had resigned because of poor health in late 1944, and South American nations insisted that Argentina be allowed to join the United Nations. The recognition of Argentina represented a victory for such officials as Assistant Secretary of State Nelson Rockefeller, who wanted the United States to preserve a regional bloc of support as it entered the United Nations. But, by the end of 1945, the department once again changed its policy toward Argentina. The new ambassador to Argentina, Spruille Braden, interfered in a hostile manner in Argentine politics. Braden, who had established a reputation as a tough foe of fascism during his wartime ambassadorships in Colombia and Cuba, denounced the military rulers of Argentina as erstwhile sympathizers of Nazi Germany. For his efforts, Braden reaped the disdain of Argentine military men, favorable press commentary in the United States, and a promotion in December 1945 to assistant secretary of state for Latin American Affairs from the new secretary of states, James F. Byrnes. From Washington, Braden continued his campaign, particularly against the leading presidential candidate, Colonel Juan Perón. Two weeks before the Argentine election, the department published the "Blue Book," a study that purported to document the fascist proclivities of prominent Argentinians. Perón cleverly seized upon Braden's impolitic behavior and turned the election into a campaign against Yankee imperialism. He won a landslide victory on February 24, 1946.[13]

In view of the stormy events that surrounded Colonel Perón's election, Byrnes and Braden were disposed in 1945–1946 to support AD, a party whose platform was prodemocratic and vigorously antifascist. Moreover, Venezuela merited their support, because it generally approved of the "Blue Book" and was one of the few Latin American nations to back a State Department scheme to promote democracy in Latin America. In October 1945, Eduardo Rodríguez Larreta, the foreign minister of Uruguay, one of Latin America's few democracies, proposed that the Pan American nations consider multilateral action against any member state violating elementary human rights. Byrnes and Braden assiduously lobbied for the Larreta Doctrine, a proposal that neatly dovetailed with their Argentine policy. The proposal received, however, an indifferent reception from almost all the Latin American countries. The response was perhaps predictable, since repressive, authoritarian regimes wielded power in most of these lands. Only Uruguay, the original sponsor, and the Venezuelan junta unequivocally endorsed the Larreta Doctrine. Upon seizing power, the junta declared, in a flush of revolutionary fervor, its opposition to totalitarian regimes and then severed relations with two notorious dictatorships, Spain and the Dominican Republic.[14] Caracas' foreign policy had the effect, of course, of securing the friendship of the two embattled State Department officials, Byrnes and Braden.

By mid-1947, both Byrnes and Braden had left Washington, and the State Department once again reassessed its Argentine policy. In January 1947, President Truman asked Byrnes to resign in a dispute over European questions and appointed General George C. Marshall, the former chief of staff and special envoy to China, to secretary of state. In June 1947, Marshall fired Braden, for he found that his Argentine policy had not weakened Perón and that it endangered hemispheric solidarity by delaying the scheduled inter-American defense conference. The United States had promised in March 1945, in the Act of Chapultepec, to form a military alliance with Latin America. Braden had twice postponed the conference because of the feud with Perón. In view of the on-going confrontations with the Soviet Union over Europe and the Middle East, Marshall decided that the goal of a united hemisphere far outweighed any misgivings about President Perón's past associations and beliefs. The new secretary concluded a regional defense treaty with Latin America at Rio de Janiero in September 1947.[15] The department's shift in concern from fascism to communism did not alter its Venezuelan policy, for AD was also anticommunist. The *adecos* often regarded the Venezuelan communists as their chief political rivals, for the two parties had waged bitter struggles for the allegiance of the oilworkers. AD leaders took special care to ensure that the State Department understood the party's position. In March 1947, for example, junta President Betancourt told Ambassador Corrigan that he supported the Truman Doctrine "100 per cent" and that "in the event of any trouble with Russia . . . the first thing I would do in such a case would be to imprison all the active Communists and treat them as the fifth columnists that they are."[16] The United States was further encouraged when, on the eve of the Berlin blockade, Venezuelan officials told the *Oil and Gas Journal* that "you may rest assured that in any future emergency our resources will be at the full disposal of those who fight for freedom and democracy."[17]

Despite changes in both personnel and policies between 1945 and 1948, the Department of State consistently supported AD. The Venezuelan party had instituted democratic national elections and it aligned its foreign policy with the United States. Yet, though professing common democratic ideals, Venezuela and the United States bickered during the *trienio*. Besides wanting a democratic Venezuela, the *adecos* had a comprehensive program to reform Venezuela's economic and social life. Its domestic programs, however, impinged upon the interests of private U.S. investors and, in particular, ran counter to U.S. foreign economic policies.

Although the junta had pledged immediately after the October 18 rebellion to respect the petroleum code and foreign investments, it was perhaps inevitable that the *adecos* would attempt to expand the nation's control over the oil industry. AD had built its reputation on critical analyses of the oil industry's role in Venezuela, and the party's chief constituency was the oilworkers. Trouble seemed imminent when the junta appointed Juan Pablo Pérez Alfonzo, the acerbic critic of the oil law of 1943, as minister of fomento (development). Pérez Alfonzo honored, however, the 1943 law. As had the López Contreras and Medina regimes, the *adecos* conceded that the nation's financial solvency depended upon the foreign oil companies. Oil taxes now accounted for over 60 percent of the government's income. As Betancourt later observed, "it would have been a suicidal leap into space to nationalize the industry by decree."[18]

On December 31, 1945, the junta ordered an extraordinary levy of Bs. 89 million ($26.5 million) on the oil companies. The junta charged that the Medina government had manipulated revenue figures in boasting that it received 50 percent of the oil companies' profits in 1944. The levy, Decree 112, was designed to equalize the companies' profits and the nation's oil income for 1945. Though Decree 112 came as a surprise and would be costly, the three major oil companies, Standard of New Jersey, Gulf, and Shell, protested only mildly. The companies knew that the oil code intended to equalize Venezuela's oil income and the companies' profits; the question was whether the law mandated a division of profits each year or over the life of the concessions. In any case, with business booming, the major oil companies could afford the extra tax. With the new concessions obtained under the 1943 law, the companies had dramatically expanded Venezuela's production during the last two years of World War II. Standard Oil's Venezuelan subsidiary, Creole Petroleum, earned, for example, over $90 million in 1945, even after it paid an extra $18 million in taxes under Decree 112.[19]

The reaction of some of the other oil companies operating in Venezuela to Decree 112 underscored both the big three's moderation and their enviable economic positions in Venezuela. One company, Pantepec Oil, represented by the influential Wall Street lawyer and former assistant secretary of war, John J. McCloy, unsuccessfully beseeched the State Department to protest the decree as an unconstitutional assertion of Venezuelan law. The department, which had been advised by Arthur Proudfit, the general manager of Creole, that "no reasonable objection could be raised to an upward revision of the income tax structure," told McCloy that it could not intercede in a Venezuelan domestic matter and advised Pantepec to petition the junta.[20] The established companies were sanguine about Decree 112, while Pantepec and others, such as Texaco and Socony Vacuum, were distressed, because of the oil law of 1943. By allowing Gulf, Standard, and Shell to renew their old concessions and purchase new ones, the law confirmed the hegemony of the three companies in Venezuela's oilfields. In the opinion of the other companies, who together produced approximately 5 percent of Venezuela's production, the big three conciliated the junta because they were "profit-fat."[21]

The established companies and the smaller ones similarly broke over the issues of collective bargaining and pay raises for the oilworkers. At the urging of the junta, the oil companies raised wages from 35 to 50 percent in 1946. This gave oilworkers salaries and benefits averaging between $4,000 and $5,000 a year, which was comparable to compensation in the United States. After the larger companies had settled, the smaller, less successful companies were forced to accept the same pay scales. Once again, the big three decided that cooperation would enhance their long-term interests by winning the goodwill of the junta and ensuring labor peace for two years.[22] The settlement was also expected to calm Venezuela's political life by wooing the oilworkers away from the communists, who opposed the *adecos*. Any revolutionary activity could only restrict the flow of oil and profits, and the companies particularly feared that the communists might impose stringent controls on the industry if they ever gained power. Thus, the companies continually aided the junta and AD. Arthur Proudfit went to Washington in mid-1946, for example, to urge the shipment to Venezuela of materials for road construction to help create jobs, for, in Proudfit's view, the junta's failure to enact promised programs "might result in a revolution instigated by certain communistic elements in the country, which will produce complete chaos."[23] The companies also invested in the Basic Economy Corporation, an agency that funded development projects, such as milk pasteurization plants and fisheries. The corporation, which was founded by Nelson Rockefeller, was designed to help the junta reach its goal of making Venezuela self-sufficient in food production.[24]

In their responses to Decree 112 and the wage demands of the oilworkers and in their willingness to aid domestic reform programs, the established oil companies, particularly Standard Oil of New

Jersey, demonstrated a degree of cooperation and farsightedness that had not always characterized their policies in the previous three decades. The harsh lessons of the Bolivian and Mexican expropriations and the events that surrounded the oil law of 1943 had forced the companies to abandon their opposition to any proposed change in their contracts. In any case, companies like Standard Oil needed work in Venezuela, for in the immediate postwar years the South American nation remained, outside the United States, the best source of oil and profits in the world. But, by the late 1950s, Standard and other oil giants would be pumping vast quantities of oil from Middle Eastern fields (see Appendix, Table C). Then the companies would be able to counter any Venezuelan demand for more taxes with the threat to replace Venezuelan oil with Saudi Arabian or Iranian crude.

Whatever the reasons for the oil companies' new stance, what they had to concede to the junta and Minister of Fomento Pérez Alfonzo was not onerous. Decree 112 and the pay raises reduced the companies' profits but did not challenge their power. And even those measures did not financially squeeze the companies. Venezuela had an oil boom during the *trienio* that rivaled the fantastic growth of the 1920s. Venezuelan wells, which produced 14 percent of the world's output, helped fuel the postwar expansion of the U.S. economy and the reconstruction of Europe. In 1948, the Venezuelan oil companies produced 490 million barrels and earned a record Bs. 1,060 million ($316 million). Creole, which produced 56 percent of Venezuela's oil, earned more in 1948 than did United States Steel.[25] With such lucrative business, the oil companies could afford moderate, conciliatory policies.

A serious confrontation between government and industry, as would take place in the 1960s, might have occurred if Pérez Alfonzo had carried out all his proposed reforms, for Pérez Alfonzo wanted to free Venezuela from the foreign oil companies. Although agreeing with other *adecos* that Venezuela was unprepared to operate the oil business, he believed that the nation should learn by either establishing mixed government-business enterprises or chartering a wholly state-owned corporation. His major criticism of the oil code was that it left Venezuela's economic destiny entirely in the hands of foreigners. Accordingly, in 1946, the minister decided that the nation would exercise its option to take its royalties in kind rather than in taxes. With its royalty oil, Venezuela could make private deals and thereby gain marketing experience as well as be in a position to assess whether the foreign companies sold Venezuela's oil at the highest possible prices. The venture proved successful, for, after Venezuela concluded some small bilateral sales, the companies, fearing the new competition, agreed in 1948 to repurchase the royalty oil at nineteen cents a barrel above prevailing prices. Pérez Alfonzo followed this step by announcing a no-new-concessions policy and appointing, in March 1948, a special commission to study the feasibility of a national oil company. These policies did not, however, immediately threaten the established oil companies' standing. Under the law of 1943, the companies had obtained over twelve million hectares of concessions, or about 9 percent of Venezuela; they did not need to lease new holdings for years. And in November 1948, the military overthrew the Gallegos government, before the special commission completed its report. More than a decade passed before another government again proposed to enter the oil business.[26]

While Venezuela and the oil companies negotiated the issues of taxes, wages, and prices, the State Department watched from the sidelines. The department initially responded to Decree 112 with "shocked surprise" and predicted that it would be a blow to the junta's standing abroad.[27] But the established oil companies decided that they could coexist with the *adecos* and refrained from requesting State Department assistance. Indeed, the department's only role in the government-industry negotiations was to explain Venezuela's tax and wage policies within the Truman administration. Throughout 1946, Secretary of War Robert Patterson prodded diplomats to safeguard U.S. oil interests. Patterson had received a warning from an excitable military attaché in Caracas that "the demands of the Venezuelan Communist-dominated oilworkers . . . will ultimately force the companies out of business or effectively curtail their control of production."[28] The department assured Patterson that it adequately protected the oil companies and noted that neither the United States nor the oil companies would gain if the companies "were encouraged to resist the reasonable demands of labor." As to the issue of communism, the department generally held throughout the *trienio* that the Venezuelan communists were politically weak, that the *adecos* would prevent communism, and that the Soviet Union exerted little influence in Venezuela.[29]

In defending the *adecos* and their labor policies, the State Department hoped to prevent what the War Department feared might happen. As during World War II, U.S. diplomats and military planners assigned strategic priority to Venezuela's petroleum. Since the mid-1930s, the rate of new discoveries of petroleum in the United States had declined; its citizens produced more petroleum annually than they discovered in new reserves. And the oil demands of the war had drained valuable supplies. After 1947, the United States would be a net importer of petroleum (see Appendix, Table D). Military experts predicted in 1947 and 1948 that the United States would fall short at least two million barrels of petroleum a day if it engaged in war with the Soviet Union. Immense quantities of reserves lay in the Persian Gulf area, but military men worried that the area was vulnerable to a Soviet attack. In case of war, the United States might have to draw on the oil reserves of the entire Western Hemisphere: Canada, Colombia, Mexico, and, particularly, Venezuela.[30] As such, the United States needed stability in Venezuela; either a civil war or a heated controversy between the government and industry could jeopardize oil production. The Department of State was pleased that the established oil companies could accommodate the Venezuelan government, and it tried to foster political calm in the nation by supporting the *adecos*.[31]

The "50-50" tax split and the pay-raise issues did not produce a confrontation between the United States and Venezuela, because the oil companies accepted the changes and because the State Department did not want to impede the expansion of Venezuela's petroleum production. Moreover, the department judged the oil policies as moderate and reasonable. Indeed, except for the sale of royalty oil and the aborted plan for a national oil company, Pérez Alfonzo's reforms seemed to differ only in degree from those of his predecessors; he was collecting taxes more efficiently than had either the López Contreras or Medina regimes. While the *adecos* were unprepared to restructure immediately the cornerstone of the economy, the foreign oil industry, they hoped to funnel the oil income into Venezuela's sagging agricultural sectors and into industrial development projects. Their blueprint for Venezuela's economic transformation clashed with the Truman administration's desire for Latin American economies oriented toward the export of raw materials.

A broad plank of AD's platform was the assertion that the state was responsible for the nation's economic welfare. Because of its substantial income, the Venezuelan government, "more than in other Latin American countries," could therefore exercise a "determining influence" in economic planning and development. In the *adecos'* view, economic planning was vital, for Venezuela was a "semicolonial, semifeudal country, a country tied to economic, fiscal and political imperialism, with an economy predominantly rural, chained by *latifundismo* . . ."[32] Specifically, Venezuela still had a lopsided, disoriented economy. Two out of three Venezuelans worked in agriculture; yet, Venezuela imported 40 percent of its food. Food prices were exorbitantly high for a poor country. An urban laborer spent as much as 60 percent of his income for his family's food needs. Moreover, old agricultural patterns persisted with a few controlling most of the land and the rest earning less than one dollar a day. The production of coffee and cacao, the major export crops, also continued the decline which had begun in the late 1920s. The industrial side of the economy was similarly unbalanced. Oil production was the only significant industry, accounting for 98 percent of Venezuela's foreign trade and over one-quarter of its national income. But the industry employed barely 3 percent of the working population. Shantytowns ringed the oil camps where peasants waited in anguished hope for a chance to work for the foreigners. And, even though employment opportunities were limited in the cities, a portentous exodus from rural areas had ensued with life so dismal in the countryside.[33]

In May 1946, the junta created a central economic agency, the Corporación Venezolana de Fomento (CVF), to lead the attack on Venezuela's economic problems. The CVF's goals were broadly defined—to increase agricultural production and to establish new industries. The agency's funding would come from oil revenues. In form, the CVF did not radically differ from the economic policies of the López Contreras and Medina regimes; it also sought "*sembrar el petróleo*," or "sow the petroleum."[34] What distinguished the CVF and the *adecos'* economic policies was the willingness to question the feasibility of U.S. principles of free trade and private investment for Venezuela.

The Truman administration professed that all nations should have free access to trade and raw materials. This "free trade" or "Open Door" principle was the keystone of U.S. foreign economic policy during the twentieth century. The principle was proclaimed, for examples, in Wilson's "Fourteen Points," the Atlantic Charter, and, most recently, President Truman's October 27, 1945, Navy Day speech. Venezuela had generally adhered to the principle in its economic relations with the United States. It had allowed U.S. oilmen virtually unlimited access to its vital natural resource, and it had signed a reciprocal trade agreement with the United States in 1939. Yet, as post-Gómez Venezuelan governments coped with the economic chaos left by the dictator, they began to violate the free-trade principle.

A minor controversy between the United States and Venezuela during World War II illustrated some of the problems which free trade posed for a developing nation. As a way of conserving essential materials and resources, the American republics, with the blessing of the United States, established import control commissions during the war; in effect, the commissions interdicted free trade. As the end of the war neared, the State Department urged its southern allies to relax their controls, noting that "it was short-sighted policy on the part of most of the nations of the world, including the United States, in increasing tariffs and restricting trade by import and other controls after the first World War and in the early thirties which helped to bring about and prolong the recent world wide depression."[35] Despite the department's use of the lessons of history, Venezuela was initially reluctant to remove all import controls. Local pharmaceutical firms had prospered during the war and feared the return of their North American competitors. Resorting to the "infant industries" argument, they pressured the Medina government to retain some restriction on imports. In several meetings, State Department officials pressed Venezuelans to drop the restrictions, while emphasizing the virtues of free trade and the folly of protecting uncompetitive industries. While the department never threatened Venezuela, it was persistent, and it carefully monitored Venezuela's trade policies until September 1945, when President Medina removed the objectionable controls. Venezuelan businessmen believed, however, that their small firms had been sacrificed for the sake of Venezuelan-U.S. harmony.[36]

The *adecos* were less willing than the Medina administration to submit to U.S. views on international trade. Part of the *adecos'* economic revitalization plan included a thriving merchant marine. With Ecuador and Colombia, the junta founded in 1947 the Flota Mercantil Grancolombiana. Venezuela and Colombia each contributed 45 percent of the funding and Ecuador supplied the remaining 10 percent. The shipping consortium intended to carry Gran Colombian goods in Gran Colombian bottoms and to reduce the three countries' dependence on the U.S. merchant marine. The three countries also hoped that the Flota would be a first step toward creating a regional trading bloc and perhaps, as the great liberator Simón Bolívar envisioned, a united northern South America.[37]

Within months after the Flota's incorporation, the State Department was engaged in a heated debate with the consortium's two principal sponsors. Both countries wanted their purchases in the United States carried on their ships and Colombia wanted the Flota to handle all of its coffee exports. Washington immediately protested the policy as an unwarranted and dangerous violation of free commerce calculated to act inevitably "as a restraint on international trade in that it would provide a monopoly to national interests with subsequent large increases in freight rates."[38] Its protests implied that the Flota resembled the economic statism that Nazi Germany had practiced. Venezuela and Colombia angrily replied that the U.S. "shipping trust" had dominated Caribbean shipping for decades and had extracted exploitative rates for their services. As junta President Betancourt put it, Venezuela was simply asserting a traditional U.S. principle, "freedom of the seas."[39]

The dispute between the United States and Venezuela over the Flota flared only briefly. After overthrowing the Gallegos administration in November 1948, the military junta bowed to the State Department's protests about discrimination against U.S. carriers and in 1953 actually withdrew Venezuela from the consortium. Though shortlived, the spat highlighted the difficulties that small Latin American nations had in accepting U.S. trade policies. Without some form of government subsidy, a new Venezuelan business could not compete with its North American counterparts. As Betancourt observed, the principle of free commerce was impractical, for it implied "abandoning the defense of our infant industries." And, in any case, the United States violated its precepts in subsidizing its merchant marine by requiring that Export-Import Bank–financed purchases be carried in U.S. ships. Moreover, free trade could not give Venezuela the economic independence that the United States considered essential. Flota officials pointed out that the United States Merchant Marine Laws read that "it is necessary for the national defense and for the development of foreign and domestic commerce that the United States possess sufficient merchant-ships"; they wondered why their countries should not also be secure.[40] Indeed, the three countries' dependency on the United States had been especially apparent during World War II. Because of the strains of a two-front war, the United States could not always provide its southern allies with the tonnage needed to handle their trade.

While during the *trienio* Venezuelans doubted the utility of free trade for their nation, their actions could hardly have displeased Washington. U.S. businessmen were allowed to sell freely in the Venezuelan marketplace. Taking advantage of the reciprocal trade agreement of 1939, they provided the nation with almost 75 percent of its imports. And their sales were substantial. Because of the oil law of 1943 and the postwar oil boom, the *adecos* had money to spend on items like machinery for their economic development projects. In 1948, the value of U.S. exports to Venezuela was over $500,000, as compared to sales of only about $60,000 in 1939. Venezuela now ranked with Brazil, Cuba, and Mexico as the most lucrative markets in Latin America for North American entrepreneurs.[41]

The State Department was as diligent in upholding the other key principle of U.S. foreign economic policy, the efficacy of private capital investment, as it was in defending the principle of free trade. Truman administration officials consistently argued that the future prosperity of South America, a continent laden with raw materials, depended upon the infusion of private U.S. capital. They pointed out that U.S. economic expansion in the nineteenth century was spurred by western European investment and that "the great economic momentum which has been fathered in this country under our system of individual enterprise could easily expand into Latin America to an extent never before visualized." After removing barriers to foreign investment and providing safeguards against expropriation, Latin America could expect "thousands of prospective investors" to enrich their economies and societies.[42]

Latin America did not always cheerfully adopt the U.S. prescription for its economic health. At postwar inter-American conferences, Latin America called for a "Marshall Plan for Latin America" or, at least, the creation of a multilateral agency to finance industrial development programs. Latin Americans feared that if they remained exporters of raw materials, without an industrial base, they would continue to be inordinately sensitive to fluctuations in world trade. U.S. diplomats and economic specialists turned such arguments aside. In their view, a revived world economy, full employment in the United States, and the political security of the noncommunist world depended upon the reconstruction of western Europe. The U.S. treasury could afford only one industrial development plan. Latin America's role in the new world order was to sell its raw materials to a reborn Europe and to absorb surplus U.S. capital.[43]

Venezuela, with its prodigious oil production, became a showcase for private foreign investment. After the enactment of the oil code in 1943, U.S. oil companies invested heavily in Venezuela. Their investments steadily grew from approximately $300 million in 1943 to $850 million in 1950. Total U.S. investment in Venezuela by 1950 was $993 million, second only to U.S. investment in Canada.[44] The State Department repeatedly reminded Venezuelans of their good fortune. For example, the department applauded the cooperation between the oil companies and the junta in investing in government-directed development programs but insisted that "the Development Corporation [the Basic Economy Corporation] function as an auxiliary to, and not a substitute for, private capital enterprise." When it learned of plans for a national oil company, the department reacted by noting, in the words of Secretary of State Marshall, that it did "not welcome such a development . . . and would hope that Venezuela could be persuaded as to the advantages of the present system of private enterprise operations with any regulatory safeguards that might be reasonably required." Marshall instructed the embassy in Caracas to express "both surprise and concern" to the government should it decide to enter the oil business.[45] And when loose talk surfaced in the constitutional convention of 1947 in debates on provisions that would give the state the right of expropriation of "any class of property," Acting Secretary of State Dean Acheson expressed his misgivings to Minister Pérez Alfonzo at a Washington meeting. As Acheson had earlier telegraphed the embassy, the "question of type of Constitution for Venezuela to adopt is clearly matter for that government alone to decide, but we believe we would be remiss in not expressing in the most friendly and sympathetic terms our apprehensions over course events seem to be taking."[46]

The Department of State's fears about Venezuela's attitude toward private foreign investment were premature by at least a decade. Despite its proclaimed goal of an economically independent Venezuela, the ruling Acción Democrática party knew it needed foreign investment. Unlike other developing countries, Venezuela owned surplus capital because of its oil income. But it sorely lacked industrial skills and technology. During the *trienio*, Minister Pérez Alfonzo invited United States Steel and Bethlehem Steel to bring their expertise to Venezuela to mine the recently discovered hills of high-grade iron ore near Ciudad Bolívar, in the eastern section of the country. The scholarly Pérez Alfonzo opposed, in principle, foreign control of extractive industries; he could not forget the oil companies' plunder of Venezuela during the Gómez years. But he also understood that without the steel companies' aid, Venezuela would lose a potentially valuable income. Pérez Alfonzo resigned himself to signing contracts, modeled after the oil code, that would give the nation 50 percent of the steel companies' profits.[47] Such policies led the department to conclude, in a mid-1948 review of Venezuelan-U.S. relations that, while many of the *adecos* "are believed to harbor the socialists' concept of government," the Gallegos administration "has proven to be as satisfactory as any other Latin American government in its relations with foreign interests."[48]

While Pérez Alfonzo based his criticisms and fears of foreign investment on the oil industry's past performance, he was developing during the *trienio* a critique of foreign investment that would crystallize into a new oil policy when he and the *adecos* returned to power in 1958, after ten years of exile. He held that the law of supply and demand, the pricing mechanism interwoven with the principles of free trade and investment, despoiled poor, raw-material exporting countries. With their sophisticated technology and engineering skill, the foreign oil companies produced abundant and, therefore, inexpensive quantities of petroleum from Venezuela's rich oilfields. With the companies selling over 70 percent of Venezuela's production to North America and Europe, low petroleum prices primarily benefited the Western industrial nations, not the Venezuelan treasury. Moreover, as a nonrenewable resource and as the world's most vital source of energy, oil had, in Pérez Alfonzo's view, an "intrinsic" value superior to its market price. What was happening was the rapid sale and depletion of a precious natural resource at an inexpensive price of approximately two dollars a barrel. Similar fates awaited other Latin American exporters of raw materials, albeit their copper, tin, and silver had less intrinsic value than petroleum. Pérez Alfonzo reasoned that petroleum-exporting countries must unite to conserve their resources and to influence prices.[49] From such formulations would spring in 1960 the Organization of Petroleum Exporting Countries (OPEC), the brainchild of Juan Pablo Pérez Alfonzo.

Though Pérez Alfonzo and his party concluded that free trade and private foreign investment hurt the underdeveloped, raw-material-exporting country, the *adecos* could only question the foreign economic policies of the United States; they could not yet challenge them. Since Venezuela lacked the requisite technology and engineering skill to operate and expand the extractive industries, any policy that disrupted foreign investment threatened to plunge the nation into an economic crisis. And the *adecos* did not have the political support to sustain their policy initiatives, such as the Flota Mercantil Grancolombiana and the national oil company. After overthrowing President Gallegos in November 1948, the military rulers displayed a conservative capitalist bent, scrapping the *adecos'* programs and realigning Venezuela with U.S. foreign economic policies. They praised and promoted Venezuelan-U.S. trade, allowed the oil companies to expand production, and encouraged future foreign investment by selling new concessions to both the oil and steel companies.

The failure to alter Venezuela's place in the world economy was only one of the *adecos'* disappointments. On the afternoon of November 24, 1948, they lost their power and their greatest achievement, a constitutional, democratic Venezuelan state. In a brief, almost bloodless action, a *golpe frío*, the *adecos* were removed by the same young military officers who put them in power. A new junta, composed exclusively of military officers, assumed control of Venezuela. The military leaders suspended constitutional guarantees, dissolved Congress and state legislatures, and discarded the new constitution. The new junta emphasized its provisional status and promised another constitution and free elections. But the aftermath of the *golpe* of November 24 would not be another try for democracy but rather ten years of military dictatorship.

The November 24 counterrevolution succeeded because the ruling AD party alienated powerful segments of the Venezuelan polity. As could be expected in a traditional, oligarchic state, the *adecos'* plans for social justice met determined opposition. On October 18, 1948, they enacted an agrarian reform law, the redistribution of land, that threatened the hegemony of the landed elite. The party's sponsorship of labor cost local manufacturers higher wages. And its desire to expand educational opportunities impinged upon the traditional prerogatives of the Catholic church. Besides antagonizing elite groups with their programs, the *adecos* made several tactical blunders. They prosecuted 168 officials of the López Contreras and Medina regimes for embezzlement and self-enrichment in office. Whatever the merits of the cases, the tribunals smacked of vendetta; the *adecos* appeared to want to destroy their opponents politically. Moreover, they foolishly refused to share power with potential allies. Because of their control of the junta, superb organizational efforts, and the credits they accrued for extending land and the franchise to the peasants, Acción Democrática won a series of impressive electoral victories, usually garnering over 70 percent of the vote. With those mandates, the *adecos* established virtual one-party rule. They apparently did not understand that the sharing of political power with other democratically oriented parties, such as the new Christian Democratic party (COPEI), might help legitimize their political and economic programs. The *adecos* became vulnerable to the enemies' charges that they used their power purely for the party's benefit.[50]

Despite the hostility of the elites, the Gallegos government could have survived if it had retained the loyalty of the armed forces, for the military was the one traditional power group that had been courted by the government. The *adecos* had increased Venezuela's military spending over 300 percent during the *trienio*; they hoped the military would pursue its professional goals and stay out of the political arena. Initially, the *adecos* succeeded. In an open letter to López Contreras, the leading officers declared that "in only eight months, the Revolution has done more for the army than your government, in which we of the Armed Forces had so many hopes, hopes that were ultimately disappointed."[51] But by 1947–1948, the young officers heard the complaints of landowners, businessmen, and clergy. They gradually concluded that the *adecos* were destroying the national unity and purpose that they thought they had instituted in October 1945. And they now judged their civilian allies incompetent and perhaps dishonest. President Gallegos, a principled man, proved to be an inept politician and administrator, who had difficulty supervising his subordinates. Beyond restoring efficient, honest rule, some of the officers, such as Chief of Staff Marcos Pérez Jiménez, perhaps hoped to enhance their political fortunes with another *golpe*. In any case, what happened on November 24, as one historian has explained, was "that a dissatisfied military, with a tradition of political intervention, overturned a government supported by the majority of the people."[52]

The Department of State, aided by incisive political reporting by its field officers, monitored the crisis. In view of its "principal diplomatic objective in Venezuela" of keeping the oil flowing, the department grew anxious over the possibility of insurgency by the political right. Diplomats reasoned that the *adecos*, if imperiled, might turn to Venezuela's small communist party; the effect would be "to drive AD to the left," perhaps leading to a revision of the oil code or even nationalization. While it did not intrude directly into Venezuela's political life, Washington tried to help the Gallegos administration during its last months. In July, the White House hosted President Gallegos, and soon after the Department of Commerce granted Venezuela special allocations of iron and steel, both of which were in critically short supply, for public works projects.[53]

On November 29, 1948, the military junta asked for the diplomatic recognition of the United States. Lieutenant Colonel Carlos Delgado Chalbaud, the nominal head of the junta, promised Walter Donnelly, the new United States ambassador, that the junta would oppose communism, vote with the United States at the United Nations, retain U.S. military missions, and respect foreign oil and iron-ore concessions. It would also consider leasing new concessions to foreign corporations. Ambassador Donnelly, noting that the junta had the allegiance of the military and landowning, commercial, and other conservative groups, recommended recognition, although it "should be neither too hastily done nor too long delayed."[54]

The *golpe* and the junta's request for recognition prompted the State Department to study its recognition policy. Officials were dismayed by an antidemocratic trend sweeping Latin America. Throughout 1948, political violence rocked Colombia. And, shortly before the Venezuelan *golpe,* the Peruvian military overthrew President José Luis Bustamente and outlawed his democratic-leftist supporters, Partido del Pueblo, the *apristas.* George Kennan, the director of the Policy Planning Staff, prepared a lengthy memorandum on Venezuela. He lamented the ouster of Gallegos but concluded that a nonrecognition policy would be futile. He recommended that at the time of recognition, the United States publicly "deplore the use of force as an instrument of political change." High Department of State Officials accepted Kennan's advice. On January 13, 1949, Acting Secretary of State Robert Lovett announced the department's decision to resume relations with Venezuela, after he took note of the junta's pledge to hold free elections. Lovett's announcement was preceded by a department press release in December that incorporated Kennan's statement on the use of force. What qualms diplomats had about establishing relations were soothed by the knowledge that both the army and the National Security Council judged Venezuela's petroleum vital to the defense of the United States.[55] That consideration and the growing concern about the spread of international communism would soon make Washington even more circumspect about questioning military rule in Venezuela.

For the oil companies, the November *golpe de estado* had little immediate effect. As it had promised Ambassador Donnelly, the junta did not interfere in the normal conduct of the oil business. Some historians have assumed that because of such policies as Decree 112, pay raises, no new concessions, and the national oil company, the foreign oil industry welcomed the overthrow of President Gallegos and Minister Pérez Alfonzo.[56] Indeed, immediately after the *golpe,* the distraught Gallegos wildly charged that the military insurgents had been allied with "foreign capital and petroleum interests."[57] But the companies had little to gain from a counterrevolution. They had developed a stable, prosperous relationship with the AD-dominated government; a civil war would disrupt production and curtail profits. They also knew that the *adecos* dare not challenge their control of production. The oil companies had openly expressed confidence in their future in Venezuela by rapidly expanding their investments during the *trienio.* The only apparent problem was Pérez Alfonzo's refusal to lease new concessions, and that was a problem for the future. The Venezuelan's ideas on prices, production levels, and trade relationships were too inchoate to be perceived by the companies as threatening.[58] To be sure, the companies would enjoy military rule, for the military rulers did not propose any changes in existing oil policy, and in the mid-1950s they leased new concessions to the companies.

During the *trienio*, Acción Democrática's attitude toward both the Department of State and the foreign oil companies was occasionally contentious, but never contemptuous. The *adecos* knew Venezuela's place in the inter-American political system and in the U.S. economic orbit. If they had respected the power of Venezuela's traditional elites as much as they respected U.S. power, the *adecos* perhaps would not have suffered a decade of exile.

6. Cold War Policies, 1949–1958

Between 1949 and 1958, U.S. relations with Venezuela were harmonious and economically beneficial to U.S. businessmen. Reverting to the practices of the Gómez years, the Venezuelan government pampered the foreign oil companies by reducing their taxes and issuing them new concessions. Venezuelan leaders also pledged their country's allegiance to the U.S.-led global struggle against "international communism." In return, U.S. officials were fulsome in their praise for Venezuela, citing to other Latin American nations its political stability and economic growth as an outgrowth of Venezuela's solicitous treatment of U.S. investors. But this mutual admiration lasted only a decade. Following the overthrow of President Marcos Pérez Jiménez in January 1958, a wave of anti-U.S. sentiment swept over the nation. Venezuelans condemned the United States for its support of a sordid military dictatorship and charged that its economic policies had kept them impoverished and dependent upon a changeable world economy. Faced with such hostility, the United States began to reassess its policy toward Venezuela and, indeed, the rest of Latin America.

From 1948 to 1958, military officers governed Venezuela. But, until late 1952, they pretended that they would restore civilian rule to the nation. After the Gallegos administration was ousted in November 1948, a military junta, headed by Lieutenant Colonel Carlos Delgado Chalbaud, assumed control of the government. Delgado Chalbaud, who was a career officer, former minister of defense, and member of an influential Caracas family, repeatedly assured the Venezuelans that free, open elections would be held once order was achieved. In effect, the Colonel meant that he and his fellow officers would govern by decree until the Acción Democrática party was eliminated from Venezuela's political life. Constitutional guarantees remained suspended through 1949 and 1950 since the junta alleged that the outlawed *adecos* still engaged in subversive activities. Others argued, however, that Delgado Chalbaud delayed elections because he was expanding his political base and waiting for an opportune time to run as a civilian candidate for president.[1]

Whatever Delgado Chalbaud's plans, they were halted by his murder on November 13, 1950. Captured on his way to the presidential palace from his home, Delgado Chalbaud was taken to an unoccupied house in Caracas and then shot. Rafael Simón Urbina, a violent, unstable man who had once launched an unsuccessful revolution against the Gómez regime, led the more than twenty assassins. Why Urbina conspired against the Colonel's life remains a mystery, for, after Urbina was captured, he also was shot by prison guards when he purportedly attempted to escape. While the motives and circumstances surrounding Delgado Chalbaud's death have never been fully explained, suspicion has always centered on the second-ranking member of the junta and chief of staff, Lieutenant Colonel Marcos Pérez Jiménez. Between 1949 and 1950, rumors abounded in Caracas that Pérez Jiménez was plotting against his superior. But no investigation has ever directly linked Pérez Jiménez to the murder.[2]

Though Delgado Chalbaud's death left Pérez Jiménez as the most powerful man in Venezuela, particularly since the army was loyal to him, he did not grasp the presidency of the junta. Perhaps conscious of appearances and unwilling to lend credence to allegations that he ordered the assassination, Pérez Jiménez designated a civilian, Germán Suárez Flamerich, as president. But military rule continued, for Suárez Flamerich, a Caracas lawyer, was a mere figurehead. And constitutional guarantees remained suspended for another two years on the pretext that order had not yet been restored. Only when the officers felt confident that the public understood that military rule guaranteed stability and economic progress would they chance an election.[3]

The junta took the electoral risk on November 30, 1952. Under a new, highly restrictive electoral law, the junta certified which parties could campaign. The major parties approved were the Christian Democrats (COPEI) and the Unión Republicana Democrática (URD), a middle-class party led by Jóvito Villalba, a hero of the "Generation of '28." Seemingly assured of victory, Pérez Jiménez trumpeted the military cause under the banner of the Frente Electoral Independiente, or Independent Electoral Front. What Pérez Jiménez and his advisors misjudged, however, were the intentions of the members of the proscribed Acción Democrática party. Though denouncing the electoral laws and initially pledging to boycott the election, most *adecos* apparently voted for Villalba's party; preliminary returns showed that the military was trailing the URD. Unsettled by the voting trends and unwilling to relinquish power, the Venezuelan armed forces struck its third *golpe de estado* in seven years by dissolving the junta on December 2 and appointing Pérez Jiménez as interim president. Then, on December 13, the new government announced the final voting returns, which predictably gave victory to the Frente Electoral Independiente. When Villalba and party leaders protested the fraud, they were arrested and exiled. Finally, on April 17, 1953, the electoral fiction ran its course when the delegates, who were presumably elected on November 30, assembled and named Colonel Pérez Jiménez as constitututional president for a five-year term.[4]

The United States reacted to the four years of political charade in Venezuela with a measure of dismay that gradually turned into resignation and indifference. In formally recognizing the junta in January 1949, the Department of State had taken note of the military leaders' expressed intention to respect the constitution and schedule a democratic election. For the next year, department officials badgered the junta in private about fulfilling its pledges. Ambassador Walter J. Donnelly repeatedly urged, for example, that the junta release political prisoners. In addition, the department banned any contacts between U.S. military officers and their Venezuelan counterparts so as not to confer prestige on the regime. Such steps left the junta with the impression that the United States did not "like" it and gave department officials the confidence to predict in January 1950 that its pressure was hastening the reestablishment of constitutional guarantees in Venezuela.[5]

But by mid-1950, the State Department's optimism had faded, as it concluded, in a major review of Venezuelan-U.S. relations, that "our efforts to encourage the restoration of a greater measure of democracy have not been successful."[6] In part, the department's admission of failure reflected an increasingly sophisticated understanding of the Venezuelan political milieu. In analyzing the *golpe* of November 1948, department officials had initially decided that Acción Democrática was primarily responsible for its own problems; the *adecos* had pushed for change too rapidly and with a zeal that bordered on fanaticism. The military intervened because it wanted to restore order and moderation in the nation. The department gradually learned, however, that military leaders would insist on a major role in any civilian government. Yet, as a political officer stationed in Caracas put it, how could "the election of a centrist government friendly to the armed forces be brought about in the political climate which now prevails without the overt intervention of the armed forces in politics?"[7] This dilemma essentially explained why the military junta postponed elections for four years.

Short of massive, direct interference, Washington could not determine the course of Venezuelan politics. But what influence U.S. officials possessed, they chose to use guardedly. The State Department wanted neither to alienate a Cold War ally nor to publicize problems in Latin America, its special area of concern, to the rest of the world. In 1949, for example, when Rómulo Betancourt and Rómulo Gallegos, the exiled AD leaders, wanted through Uruguay to denounce the military rulers at the United Nations for violations of human rights, the department dissuaded the Uruguayans from accepting the request, after arguing that the airing of Western Hemisphere troubles "would play into the hands of the Soviet bloc and might produce an undignified squabble among the Latin American nations."[8] After the outbreak of the Korean War, officials became even more circumspect "in expressing our lack of enthusiasm for the present military dictatorship." By supporting the position of the United States at the United Nations, sending financial aid to Seoul, and cooperating in inter-American defense planning, the Junta Militar made it easier for the Department of State to overlook the loss of political and civil liberties in Venezuela.[9]

The United States was principally interested, of course, in Venezuela's oil. As the State Department conceded in its policy review, "all U.S. policies toward Venezuela are affected in greater or lesser degree by the objective of assuring an adequate supply of petroleum for the U.S., especially in time of war."[10] Specifically, the department believed that it must instruct Venezuelans on how to protect their oilfields from sabotage and external military threats and persuade them that continued private U.S. ownership of their oil resources was in the best interests of both Venezuela and the noncommunist world. These goals superseded whatever distaste the department had for a military dictatorship.

Fears that the on-going confrontations with the Soviet Union in Europe might erupt into world war prompted Defense and State department officials to safeguard Venezuela's oilfields, which national security managers continued to believe would be vital for victory. With the permission of the Junta Militar, a "security survey team" led by James Coulter, a former agent of the Federal Bureau of Investigation (FBI) and an expert on plant protection from sabotage and espionage, toured oil facilities in early 1949. Noting that some installations were vulnerable to sabotage, Coulter recommended heightened security. The oil companies cooperated by submitting their employees' fingerprints to the FBI to determine if they were communists or "fellow-travelers." In addition, they increased surveillance of their property and accepted the offer of the Central Intelligence Agency to watch for subversives and saboteurs among Venezuelan oilworkers.[11] The junta also responded by pledging "to move against the Communists here in the event of serious trouble either in Venezuela or in the international field." Indeed, in May 1950, the junta outlawed one wing of the nation's Communist party when its members, who allegedly comprised 10 to 20 percent of the oilworkers, went on strike to protest the military's repressive policies. Two years later, the government, charging that "the Soviet Embassy here has been a permanent center of clandestine propaganda devoted to incite domestic unrest and to disturb the public security and peace," severed relations with the Soviet Union.[12]

Beyond securing Venezuela's oilfields, national security planners wanted to protect them from "air, submarine, and surface raiders." After the outbreak of the Korean War and as part of its new stance toward the junta, the State Department allowed defense officials to approach Venezuela about joint action in case of war.[13] Colonel Delgado Chalbaud readily accepted the U.S. offer. Cooperation was facilitated by the department's decisions in 1950 and 1951 to renew its military missions in Venezuela and by congressional approval of the Mutual Security Act of 1951, which gave the executive branch the power to negotiate military assistance agreements so that, in Assistant Secretary of State Edward Miller's words, "we can exercise a great deal of leverage, to the end of having the armed forces of those countries in Latin America better allies of the country." Under the mutual security program, Venezuela would receive in the 1950s over $30 million in credits for military purchases.[14]

Curbing economic nationalism and maintaining Anglo-American control over Venezuelan resources proved as easy for the Department of State to accomplish as arranging protection for those resources. Following the overthrow of the Gallegos administration and its controversial minister of development, Juan Pablo Pérez Alfonzo, Venezuelan oil policy underwent a period of drift and uncertainty. During the *trienio*, Pérez Alfonzo's policies of higher taxes, a national oil company, and no new concessions had been sharply questioned by conservative, business-oriented groups, such as the Venezuelan Federation of Chambers and Associations of Commerce and Production (Fedecámaras). These groups, many of whom supported the junta, warned that the previous government had jeopardized the nation's economy by failing to expand the known reserves to guarantee future production. Moreover, they worried that Venezuela's tax structure would make their oil uncompetitive with Middle Eastern oil, large quantities of which were entering the world market in the early 1950s. Some even opined that production should be expanded while oil was still in demand, for in the future a new and cheaper source of energy, such as nuclear power, might leave oil economically uncompetitive.[15]

The foreign oil industry reinforced Venezuelan doubts by fiercely attacking the ideas of Pérez Alfonzo. The industry's theme, as exemplified in a series of speeches given in Venezuela by petroleum economist Joseph Pogue of Chase Manhattan Bank, was that "contrary to popular view, the key to the petroleum industry is the market, not the producing oil well." Pérez Alfonzo had succeeded in raising taxes and controlling production only because he had operated in a "seller's" market of oil shortages in the immediate postwar years. But Middle Eastern oil, though perhaps vulnerable to attack by the Soviet Union, was of better quality, cheaper to produce, and closer to European markets than Venezuelan crude. Moreover, as the world became "more amply endowed with oil" and world prices declined, Venezuela could expect oil and coal producers in the United States to push for restrictions on Venezuelan imports, as they had during the early 1930s. The consumer, Pogue warned, wanted only a reliable and cheap source of supply. If Venezuela did not wish to lose its customers, it must reverse Pérez Alfonzo's policies by lowering its taxes and labor costs and attracting investors by offering new concessions.[16]

Venezuela's oil officials essentially succumbed to the arguments of native businessmen and oil industry spokesmen. After the removal of Pérez Alfonzo, the Ministry of Fomento, under the supervision of officials who had previously served in the López Contreras and Medina regimes, conducted an extensive study of the nation's oil policies. The report, which was submitted to the junta in early 1950, recommended the leasing of new concessions and measures to control the costs of oil production.[17] Significantly, the report's submission followed an unsuccessful tour of Saudi Arabia and Iran by ministry officials. The purpose of the trip was to bring "about a better understanding amongst oil-producing countries with the ultimate objective of increasing the standard of living in those countries." But the Saudis, angered by Venezuela's support for the partition of Palestine, refused to meet the Venezuelan delegates, and the Iranians merely accepted a copy of Venezuela's oil code translated into Persian.[18] Both nations, which were about to challenge Venezuela's domination of the world oil market and reap massive new revenues, saw little reason to help Venezuela protect its production and prices.

While the military rulers accepted the Ministry of Fomento's report and even entertained bids on new oil concessions in 1951, they neither granted any new leases nor launched any other new oil policies between 1948 and 1952. The junta understood, as a confidential opinion poll commissioned by Creole Petroleum revealed, that an overwhelming majority of Venezuelans wanted not new leases and tax incentives for the foreign oil industry but rather restrictive measures leading to nationalization. Moreover, a major plank in the platform of the URD, the junta's leading opponent in the election of 1952, was essentially a restatement of Pérez Alfonzo's policies with demands for a national oil company, a ban on concessions, and higher taxes.[19] Accordingly, the junta, through its Frente Electoral Independiente, also expediently pledged to guard the nation's natural resources. Nevertheless, it was the first government since the Gómez era not to press any new demands on the foreign oil industry. Indeed, throughout the tenure of the junta, oilmen generously complimented the military rulers for their "fairness."[20]

The State Department also helped shape Venezuela's oil policies. It defined its mission as "creating a favorable atmosphere for the continuance of U.S. private ownership of the oil industry" and "preventing any widespread demand for nationalization by the Venezuelan government." This mission was crucial, since the United States was now a net importer of petroleum. The president's Materials Policy Commission, chaired by William S. Paley, highlighted this historic change in its 1952 report. Citing the increasing requirements of the United States for raw materials for national security and economic expansion, the report declared that "private investment abroad must be the major instrument for increasing production of materials abroad" for U.S. use.[21] To accomplish this goal, Washington employed a variety of diplomatic tactics. Public relations included an all-expense-paid tour of the United States for an amateur baseball team of Venezuelans and frequent public praise for the nation's cooperation with U.S. capitalists and its "Open Door" policies. These public gestures were reinforced by vigorous lobbying of Venezuelan officials, such as Minister of Fomento Manuel Egaña, to persuade them to control taxes on petroleum and grant new concessions.[22] A more tangible inducement was provided by the Department of State's decision to expand the reciprocal trade agreement of 1939 with Venezuela. The supplementary agreement, which was signed on August 28, 1952, reduced the U.S. tariff on the type of low-gravity, heavy crude oil that Venezuelan fields produced. Beyond gaining some new concessions for U.S. exporters, the new tariff eased Venezuelan fears about Middle Eastern competition and ensured that its crude would remain available to the United States in case of war. The supplementary agreement also was intended, as Assistant Secretary of State John M. Cabot later pointed out, "to safeguard important American investments."[23]

During its second term, which virtually coincided with the rule of the military junta, the Truman administration achieved most of its goals for Venezuela. Diplomatically, militarily, and economically, Venezuela remained firmly within the sphere of influence of the United States. In perhaps its last survey of Venezuelan-U.S. relations, the administration proudly concluded that Venezuela had profited, even been transformed, by this relationship. Asserting that "oil has shown the way to national prosperity," the Department of State predicted that "the success of Venezuela's national venture would appear to depend upon continued application of its liberal investment policies, particularly with respect to foreign capital, and intensification of its programs of social welfare and economic diversification." The department admitted, however, that not all of its expectations had been realized. The study, which was prepared shortly after the electoral fraud in December 1952, pointed to the lack of popular participation in the country's political life. Moreover, it noted that "a majority of the nation's population remains ill-fed, ill-clad, poorly housed, illiterate, and disease weakened." What the department seemed to suggest was that these problems would be resolved not by raising taxes or oil prices but by Venezuelan leaders demonstrating a greater concern for political and social justice.[24]

A commitment to human rights characterized neither the regime of Colonel Marcos Pérez Jiménez nor the Venezuelan policy of the administration of Dwight D. Eisenhower. With his inauguration as president in April 1953, Pérez Jiménez formally claimed the power he had been exercising since the murder of Delgado Chalbaud. The military dictator, like Generals Castro, Gómez, López Contreras, and Medina, was from Táchira, born in the town of Michelena in 1914. A graduate of the Military Academy and an outstanding student, Pérez Jiménez rose steadily through the officer ranks. After helping lead the *golpe* against President Medina in October 1945, he served as chief of the general staff during the *trienio*. Pérez Jiménez' ambition and frustration with civilian rule were evident to the leaders of Acción Democrática, for they alternately tried to placate him with new orders for military equipment or to remove him from political life by sending him on an extended tour of Latin America. Their ploys failed to dissuade the colonel, however, for he led the military conspirators who overthrew the Gallegos government in November 1948. In the military junta he served as minister of defense. A pudgy, physically unattractive man, Pérez Jiménez relied on his skills as an organizer and planner to fulfill his goals.[25]

When it defended itself, the military dictatorship pointed to its accomplishments under the "New National Ideal." While cloaked in platitudes and vaguely defined, the new philosophy put a higher premium on national unity and material and technological progress than on political freedom and intellectual and moral improvement. The armed forces, disciplined, trained, and ostensibly nonpartisan, could best carry out this mission. These ideas were born, in part, during Pérez Jiménez' experience as a young officer in Peru on a training tour. There he had met Peruvian officers, including future strong-man Manuel Odría, who believed that the military must play an expanded role in the nation's life and promote technological progress. Moreover, he had been impressed with the splendor and modern conveniences of Lima and compared Caracas unfavorably to the Peruvian capital. As dictator, Pérez Jiménez maintained close ties with Odría, and his profligate spending on Caracas was perhaps intended to impress the Peruvian.[26]

In practice, the New National Ideal consisted mainly of lavish public works projects for Caracas. The construction of new hotels, office buildings, apartments, and super highways transformed Caracas into a glittering, architecturally modernistic city, which rarely failed to impress foreign visitors. Other than on public construction, Pérez Jiménez concentrated his energies and the nation's money on the armed forces, the mainstay of his regime. Soldiers lived like aristocrats with impressive barracks and social clubs and the latest in military hardware. He also allowed them to handle public works contracts, which offered excellent opportunities for embezzlement and fraud. For the highest-ranking officers, there were infamous vacation spas, where comely prostitutes plied their trade.[27]

Along with revivifying Caracas and bestowing favors on the military, Pérez Jiménez suppressed political and civil liberties with a thoroughness reminiscent of the Gómez tyranny. An extensive secret police network, Seguridad Nacional, headed by the brutal Pedro Estrada, hunted members of Acción Democrática and the Venezuelan Communist party. Capture could mean torture and imprisonment in one of Estrada's notorious dungeons, an experience one scholar attested left individuals who could never "be restored to useful existence after spending a few days as the guest of the Seguridad

Nacional."[28] As well as banning political opposition, Pérez Jiménez outlawed other institutional bases of power. The oilworkers' union was disbanded and replaced with a confederation subservient to the government. When its suppression of unions received international criticism, the government withdrew from the International Labor Organization. And the peasant syndicates, formed during the *trienio,* were crushed and their leaders imprisoned and exiled. Indeed, the dictatorship repossessed almost all of the land acquired between 1945 and 1948 for agrarian reform and sold it to private investors and speculators.[29] In an area replete with repressive governments, one correspondent labeled the Pérez Jiménez regime "the toughest dictatorship in the hemisphere."[30]

The Eisenhower administration conducted cordial relations with the unsavory government of Pérez Jiménez. While undoubtedly preferring to work with decent, democratic regimes, the administration and Secretary of State John Foster Dulles' primary interests in Latin America were to promote private U.S. investments there and, in particular, to eradicate any communist influence. Through most of the 1950s, Latin America was assigned a low priority within the Department of State, for, except for Guatemala, it was considered beyond the grasp of the Soviet Union. Internal communist subversion might be the only problem. Since he concentrated on the global struggle against China and the Soviet Union, Secretary Dulles judged Latin American leaders on whether they kept their nations internally secure. As John Dreier, the U.S. representative to the Organization of American States reminisced, the secretary wanted to see "flourishing little democracies in Latin America" but believed "that it was not in the nature of things" and thus was "inclined to feel that governments which contributed to a stability in the area were preferable to those which introduced instability and social upheaval, which would lead to Communist penetration."[31] Dulles' support for dictatorships, as long as they were firmly anticommunist, was not a new pattern in the Latin American policy of the United States. The exigencies and assumptions of the Cold War, particularly after Korea, had led the Truman government to accommodate the military junta in Venezuela. Perhaps what distinguished its policies from the previous adminstration was the Eisenhower administration's unabashed embrace of military dictatorships and its failure to criticize, however mutely, the rampant political and civil repression in Latin America.

On diplomatic grounds, Pérez Jiménez' regime qualified for Washington's support. The government's anticommunist credentials included the continued persecution of the small Venezuelan

Communist party and a freeze on relations with the Soviet Union, although, ironically, the one independent party the government worked with was the Partido Revolucionario Communista, a coterie of Communists in Venezuela known as "the Blacks" who informed on "the Reds."[32] Venezuela also mimicked U.S. foreign policy; it was, for example, the first nation to pledge publicly its assistance during the Formosan crisis of 1955. And it constantly assured the United States that it vigilantly guarded against the communist menace, as when security chief Estrada visited Washington, where he was feted by high diplomatic and security officials, and declared "our great resources of strategic materials have been, are, and always will be, the main objective of subversive activities."[33] But Pérez Jiménez' most appreciated contribution was his role as host in 1954 of the Tenth Inter-American Conference in Caracas, where Secretary Dulles successfully lined up countries behind a resolution that denounced the "international communist movement" as "a threat to the sovereignty and political independence of the American States." The resolution presaged the Central Intelligence Agency–directed overthrow of President Jacobo Arbenz Gúzman of Guatemala, an accomplishment during its first term that the Eisenhower administration considered the signal achievement of its Latin American policy.[34]

The United States thanked Pérez Jiménez with medals and military support. During the 1950s Venezuelan soldiers frequently received tours of U.S. military facilities, instruction from U.S. military advisors stationed in Venezuela, and credits, under the Mutual Security Act, to purchase arms. The United States also agreed in 1957 to lend Venezuela approximately $180 million, over a period of ten years, to purchase military matériel and services.[35] Gifts for Pérez Jiménez included the conferring in 1954 upon the dictator by President Eisenhower of the Legion of Merit, the nation's highest honor for foreign personages. The award's announcement cited the Venezuelan's "indefatigable energy and firmness of purpose" for having "greatly increased the capacity of the Armed Forces of Venezuela to participate in the collective defense of the Western Hemisphere" as well as noting that "his constant concern toward the problem of Communist infiltration has kept his government alert to repel the threat existing against his country and the rest of the Americas." The medal was pinned on the dictator in a grand ceremony by Ambassador Fletcher Warren, a career diplomat who became a supporter and close friend of Pérez Jiménez and Estrada. Perhaps hoping to reassure Pérez Jiménez that he remained in U.S. esteem, the Navy Department named him an "honorary submariner" in 1957.[36]

Pérez Jiménez had not won the respect of the Eisenhower administration merely because he was an anticommunist crusader. As the citation for his Legion of Merit award proclaimed, "his wholesome policy in economic and financial matters has facilitated the expansion of foreign investment, his Administration thus contributing to the greater well-being of the country and the rapid development of its immense natural resources."[37] Like the Truman administration, the Republicans believed that Latin America's development depended upon foreign investment and free trade and sound monetary and fiscal policies and that progress would not be achieved through economic nationalism, state capitalism, or an overemphasis on industrialization. As with political and military relations, the difference between the Truman and Eisenhower administrations' approaches to foreign economic policy in Latin America was a matter of degree; the Republicans, led by Secretary of the Treasury George Humphrey, preached about the virtues of "enlightened private enterprise" with a zeal and stridency not heard since the era of Herbert Hoover. Moreover, they resisted all Latin American entreaties for their own Marshall Plan. Indeed, during the 1950s, the communist nation of Yugoslavia received more foreign aid from the United States than did all of Latin America.[38] As one State Department official, who was responsible for economic affairs with Latin America, disgustedly recorded in his private papers, the policies of the Eisenhower adminstration "contributed to the widespread belief in Latin America that the principal objective of American policy was to make Latin America safe for American big business."[39]

U.S. corporations were not only safe but also prospered well in Pérez Jiménez' Venezuela. During the 1950s, direct investments in the nation more than doubled, rising to approximately $2.5 billion in 1960. This represented more than 25 percent of all U.S. direct investments in Latin America. Investments in oil, nearly $2 billion, led the list with the bulk of the remainder invested by steel companies in Venezuela's rich iron-ore resources. Trade, also, grew steadily during the decade. The value of U.S. exports to Venezuela surpassed one billion dollars in 1957, making Venezuela, a nation of only seven million people, the sixth best market in the world for U.S. traders. In turn, U.S. businesses purchased 40 percent of Venezuela's oil exports, more than 70 percent of its coffee and cacao, and most of its iron ore. The reciprocal trade agreement, which granted tariff concessions on 98 percent of Venezuela's trade with the United States, ensured that the South American nation's economy remained tightly bound to the U.S. market.[40]

The return on investment and trade in Venezuela was handsome during Pérez Jiménez' rule. Between 1950 and 1957, the oil companies earned $3.79 billion dollars as profit, with earnings of $828 million in 1957, representing the best year the oil companies would ever have in Venezuela. The government allowed the companies to remit these profits and dividends without restrictions. Creole Petroleum, for example, accounted for nearly half of the dividend income of its parent, the gigantic multinational corporation, Standard Oil of New Jersey.[41] The U.S. steel companies also prospered, as they began to ship iron ore out of the country in the 1950s. And the more than 35,000 U.S. citizens who lived and worked in Venezuela appreciated the negligible income taxes. As one *yanqui* banker told a visiting correspondent, "You have the freedom here to do what you want to do with your money, and to me, that is worth all the political freedom in the world."[42]

Indeed, Pérez Jiménez' generous, permissive policies toward foreign capital virtually guaranteed prosperity for U.S. businessmen. The dictator kept a tight rein on the oilworkers, prohibiting strikes and decreeing moderate wage increases at scheduled intervals. Moreover, he imposed no new taxes on the oil and steel industries and actually collected a slightly smaller percentage of the companies' income than had Pérez Alfonzo.[43] Instead of increasing the nation's income by raising taxes as had the López Contreras, Medina, and Acción Democrática governments, Pérez Jiménez chose to boost government revenue by allowing the companies to expand production. Between 1950 and 1957, oil production rose from 547 million barrels a year to over one billion as predictions of a glut of oil on the world market proved premature; instability in Iran between 1951 and 1953 and the Suez Canal crisis of 1956 retarded the growth of production in the Middle East. What the companies most enjoyed, however, was Pérez Jiménez' decision in 1956–1957 to lease for forty years another 821,000 hectares of concessions for $675 million.[44] His action implied that the industry's future was secure and that Venezuela had discarded Pérez Alfonzo's goal of preserving the national treasure for the day when Venezuela could exploit its own oil.

In return for these concessions, the foreign companies behaved during the Pérez Jiménez era. They paid their taxes. And they worked with the government on technical improvements, such as reinjecting natural gas produced with oil back into the ground instead of wastefully flaring it. Moreover, oil executives promised to employ fewer U.S. workers and train more Venezuelans for staff and executive positions. Salary and housing distinctions were now based on occupation rather than nationality; Venezuelans had been segregated from their U.S. employers in the oil camps. Corporate citizenship also included contributions to educational and scientific foundations.[45] What perhaps failed to endear the companies to some Venezuelans was their constant and profuse praise for the regime's "forthright" and "progressive" policies. As one critic noted, "the American business and industrial community in Venezuela constituted one of the most enthusiastic cheering sections Pérez Jiménez ever had."[46]

This spirit of cooperation between government and U.S. private enterprise delighted the Eisenhower administration. Not only was the United States willing to support Venezuela but also it held the country up as a model to "underdeveloped" nations. Upon arriving in Caracas after attending in November 1954 an acrimonious inter-American economic conference in Brazil where the United States again denied requests for economic assistance, Secretary Humphrey told newsmen that he sensed "something special in the air" in Venezuela, for the nation was "an example of what can be achieved when private enterprise is stimulated in an atmosphere of economic freedom." Assistant Secretary of State Henry Holland agreed with Humphrey's assessment when, in testimony to the Senate in 1956, he stated that "Venezuela is a sort of showcase of private enterprise" and that it had a standard of living that far exceeded that of any other Latin American country. Holland's statement merely reiterated that of his superior, Secretary of State Dulles. As the secretary remarked to the Senate Committee on Finance in 1955, Venezuela was "a country which had adopted the kind of policies which we think that the other countries of South America should adopt." Dulles added that if other nations emulated Venezuela in creating "a climate which is attractive to foreign capital . . . the danger of communism in South America, of social disorder will gradually disappear."[47]

The "showcase" of U.S. policy in Latin America displayed, however, more than just the glittering edifices of Caracas. Despite the influx of foreign capital, Venezuela remained a poor nation which depended on the sale of one product for its solvency. Per capita income, while higher than most Latin American countries, was a meager $500 a year. A Venezuelan consumed on the average 1,950 calories a day as compared to an average consumption of 3,200 calories in the United States. Life expectancy at birth for the Venezuelan was at least ten years fewer than for a North American. Even during the oil boom of the 1950s, perhaps 10 percent of the population was unemployed. And a United Nations commission reported that almost half of Venezuelan adults were illiterate, which was predictable since the 5 percent of the budget annually appropriated to education under Pérez Jiménez was frequently the lowest share in the hemisphere.[48]

Many of these statistics could be ascribed to life in the countryside. Here the poorest lived with per capita incomes still less than $100 a year. A decline in agricultural output and productivity, which had plagued Venezuela since the end of World War I, continued during the 1950s, for the government neither redistributed land nor invested in rural modernization. Exports of coffee, the nation's traditional cash crop, fell to 470,000 bags in 1957, 10 percent fewer bags than the average between 1945 and 1948 and over 500,000 fewer bags than in 1939. Production of food was inadequate; the U.S. Department of Agriculture estimated that one-third of the calories consumed by the people came from imported food. Hoping for work in the booming construction industry, many landless peasants abandoned a dreary existence and moved to the city. By the end of Pérez Jiménez' regime, two out of every three Venezuelans lived in an urban area. The population of Caracas soared from 300,000 in 1945 to over one million by 1958. This urban migration was accelerated and exacerbated by a tremendous population increase, which saw the nation's population rise, both from natural increase and from European immigration, from five million to seven million during the 1950s. The new urban dwellers huddled in squalid shantytowns, or *ranchos*, which ringed the major cities. Pérez Jiménez' showy public works programs failed to provide either adequate housing, sanitary facilities, or enough work for the largely unskilled and illiterate *campesinos*.[49]

What prosperity Venezuela enjoyed depended more than ever on the foreign oil industry and world market conditions. During the oil boom of the 1950s, Venezuela became, in Rómulo Betancourt's words, "a petroleum factory." The oil industry provided the government with over half of its revenues, about 75 percent of its income taxes, and almost 95 percent of its export earnings. This is an industry that employed only about 44,000 Venezuelans, less than 2 percent of the labor force.[50] Pérez Jiménez had not pursued, as the *adecos* had planned, programs for economic diversification. And private foreign investors, preferring the highly profitable extractive industries of oil and iron ore, shunned opportunities in housing, transportation, agriculture, and labor-intensive industries. By permitting the companies to increase production and by issuing new concessions, Pérez Jiménez, like Gómez, had placed his nation's fate in the hands of the international oil companies. Through most of the 1950s, these policies seemed perspicacious because the companies readily found markets for Venezuelan crude and at a price that steadily rose; government income more than doubled during the dictatorship. But, by the end of the decade, the nation's precarious reliance on one product and the goodwill of others was evident as oil flooded world markets. Time would also expose the folly of not pushing for new taxes during the years of plenty when, for the last time, Venezuelan oil dominated the international petroleum trade.

President Pérez Jiménez would not, however, have to live and rule with the consequences of his oil policies. In January 1958, a mass uprising of Venezuelans forced the dictator and his henchmen into exile. Ten years of military dictatorship ended because outlawed political groups were able to organize resistance to the regime and attract crucial support from some of the nation's traditional elites. By the end of his rule, Pérez Jiménez had alienated virtually every political group in the nation. The *golpe* of 1948 had been aimed at Acción Democrática, but imprisonment and exile eventually came to the URD and the Communists in 1952 and the Christian Democrats in 1957. These groups, who had initially welcomed the overthrow of AD, gradually joined the clandestine resistance movement organized by Acción Democrática. Their cooperation was achieved by assurances from the *adecos* that, once the dictator was deposed, the aspirations of all political parties would be considered and there would not be a return to one-party rule as during the *trienio*. The four parties conspired in the Junta Patriótica, a front led by Fabricio Ojeda, a newspaperman and member of URD.[51]

The political dissidents tapped a deep and widespread public disgust with the dictatorship. Unlike Secretary of State Dulles, the *campesinos*, workers, and urban squatters knew that Pérez Jiménez' policies had not benefited the majority of Venezuelans. Only so much vicarious satisfaction could be gained from the president's lavish spending and personal excesses. Moreover, they had been insulted by his latest political trickery. By the mandate of his own constitution, Pérez Jiménez' presidency was scheduled to end in April 1958. Unwilling to risk another fiasco like the election of 1952, by allowing a contested presidential race, Pérez Jiménez granted his constituency a plebescite on his rule in December 1957. On December 20, the public learned that over 80 percent of those participating had cast a vote of confidence for their leader; on the same day, Venezuelans were also told that Pérez Jiménez would be president for another five-year term.[52]

Even with most political leaders and a substantial number of citizens opposing the regime, the movement to overthrow Pérez Jiménez would probably have been protracted and exceedingly violent had not some of his former supporters joined the resistance. The Roman Catholic church, which had disapproved of the secularism of

AD and received financial assistance for its schools from the dictatorship, gradually turned against the government, citing its corruption, tawdry behavior of its leading officials, and disregard for the poor. Indeed, sustained, popular protest against the government ensued after the archbishop of Caracas read from the pulpit a pastoral letter, dated May 1, 1957, which charged that "the immense majority of our people are living under subhuman conditions." Prominent businessmen also deserted Pérez Jiménez. They had become agitated by the dictator's tendency after 1956 not to pay promptly the government's debts; an accounting of the government's books after Pérez Jiménez' overthrow found $700 million in the treasury but inexplicably unpaid bills amounting to over $150 million. They were also encouraged to defect by pledges from exiled political leaders Rómulo Betancourt and Jóvito Villalba that neither planned radical socioeconomic changes when they returned. Finally, Pérez Jiménez fell from power quickly because he lost the unqualified support of the military. Despite the privileges they received from the regime, not all members of the armed forces were blindly loyal to Pérez Jiménez. Junior officers feared that the president, by placing sycophants in key positions, was abusing the institutional integrity of the military. Moreover, they were jealous of Pérez Jiménez' growing inclination to rely on the Seguridad Nacional, rather than the armed forces, for advice on governmental affairs.[53]

The beginning of the end for Pérez Jiménez came on the morning of January 1, 1958, when the air force rebelled. Though troops loyal to the dictator suppressed the rebellion, the outbreak ushered in three weeks of antiregime manifestos, demonstrations, and strikes by civilians organized by the Junta Patriótica. Clashes with police left several hundred Venezuelans dead and over one thousand wounded. Faced with mounting civilian and military resistance, General Marcos Pérez Jiménez fled to the Dominican Republic and the arms of fellow dictator Rafael Trujillo.[54]

Rear Admiral Wolfgang Larrazábal, commander of the navy, assumed control of an all-military junta on January 23, 1958. Public protests induced Larrazábal, however, to include civilians in his provisional government. These and other demonstrations combined with a determination by the major political parties, born out of the experience of the *trienio*, to temper their differences and maintain civilian unity were crucial in thwarting attempts by some military officers and followers of Pérez Jiménez to reinstitute authoritarian rule. Larrazábal was also adept at keeping the armed forces under control and rallying those who wanted a restoration of civilian rule within democratic institutions. A tumultuous political year ended with the election of Rómulo Betancourt as president on December 7, 1958, in a fair and open contest. The leader of Acción Democrática was inaugurated president of Venezuela on February 13, 1959.[55]

Reversing what now seemed insensitive and even counterproductive policies, the Eisenhower administration, after some hesitation, responded favorably to the *golpe* of January 23 and subsequent political developments in Venezuela. Until January 1958, the United States maintained warm relations with Pérez Jiménez. Apparently, only once had the United States become exasperated with the Venezuelan tyrant. Then, in 1956, at a meeting of the American presidents in Panama, Pérez Jiménez had embarrassed Eisenhower by proposing an inter-American lending agency and then flamboyantly pledging Bs. 100,000,000 ($30 million) to the project "provided that the other American states using the Venezuelan contribution as a basis, given in proportion to their executive budgets." Eisenhower deflected this challenge to contribute nearly 4 percent of the federal budget to economic aid for Latin America by lamely suggesting the formation of a committee to study ways to improve "the welfare of the individual." Despite this insult, the administration supported its "honorary submariner" until his political demise. And in March 1958, it permitted Pérez Jiménez and security chief Estrada to enter the United States.[56]

But, by the end of 1958, the Eisenhower government had developed a new attitude toward the Venezuelan political milieu and, indeed, was reassessing its Latin American policy partly because of its experiences with Venezuela. The Department of State's first public reaction to the overthrow of Pérez Jiménez was the cautious statement in late February that "while we are not in a position to intervene in the internal developments of the countries of Latin America, we are in a position to feel, and we do feel, satisfaction and pleasure when the people of any country determinedly choose the road to democracy and freedom."[57] This observation and the recognition of the provisional government came after the department and the oil companies had received public pledges from Admiral Larrazábal and private assurances from Betancourt that foreign investment would be respected.[58]

A critical review of Washington's past policies toward Venezuela awaited, however, the concern that followed the near killing of Vice-President Richard M. Nixon in Caracas in May 1958. The vice-president, who was concluding a troubled tour of Latin America, was attacked in his limousine on the way to the capital from the airport by a howling mob angered by the United States' past connivance with Pérez Jiménez and its present harboring of the fugitive dictator.[59] The ugly attack on Nixon shocked the United States; relations with the model Latin American country were in a shambles. As Samuel Waugh, the president of the Export-Import Bank and an official who accompanied Nixon, noted in a letter to his family: "The Caracas incident was disgraceful in every sense of the word. I must confess that before we landed in Venezuela I thought, based on our nearly one month visit four years ago, that here was a country where we would be well received. You know, of course, that American capital has spent millions of dollars in developing the oil and mineral resources of Venezuela and the country has been one of our best customers for years. Probably more students from Venezuela are in the United States than from any other country in Latin America. I couldn't have been more wrong."[60]

The attack in Caracas was only one, albeit the most dangerous, of a series of confrontations for Vice-President Nixon during his tour of Latin America. He found Latin Americans bitter over U.S. policies, being told that his government backed regimes that blocked social change. He also heard the familiar theme that the United States had "neglected" Latin America since the end of World War II. These pleas for economic assistance now had, however, a special urgency, for world market conditions had been unfavorable for most of Latin America during the 1950s. On a per capita basis, the value of their exports, except oil, actually declined during the decade and the prices of their raw-material and primary-product exports fluctuated wildly. The price of coffee, for example, fell 60 percent. Eleven Latin American countries had per capita incomes of less than $200.[61]

The unhappy results of Vice-President Nixon's tour of Latin America added weight to the arguments of domestic critics of Eisenhower's Latin American policies. Within the government, the president's brother, Milton Eisenhower, and the new under secretary of state for economic affairs, C. Douglas Dillon, challenged the prevailing wisdom that infusions of private capital alone could solve Latin America's pressing social problems.[62] Congressional critics, such as Senators Wayne Morse of Oregon, J. William Fulbright of Arkansas, and Frank Church of Idaho, sharply questioned administration witnesses on why most aid to Latin America had been for internal security rather than for economic development.[63] Influential businessmen, such as Harry Guggenheim and Lamar Fleming of Anderson, Clayton and Company, also expressed their concern, urging the government to pay more attention to Latin America. What they feared was that the growing hostility toward the United States, which they considered unprecedented, might jeopardize U.S. trade and investment in Latin America.[64]

Stung by this domestic and international criticism, the Department of State began to express its preference for political democracy and respect for human rights. In effect, it began to heed Nixon's suggestion of "a formal handshake for dictators; an *embraso* for leaders in freedom."[65] The new approach seemed advantageous, for between 1956 and 1960, ten military dictators in Latin America fell from power. In August 1958, for example, President Eisenhower heartily welcomed the new Venezuelan ambassador and announced that "authoritarianism and autocracy of whatever form are incompatible with ideals of our great leaders of the past." The department, through the Immigration Service, also initiated proceedings, a process that lasted four years, to declare Pérez Jiménez an undesirable alien.[66]

Expressions of enthusiasm for representative governments, while a break from past policies, cost the United States nothing. What also marked Washington's new look toward Latin America was a willingness to spend public money in the region. By the end of the Eisenhower administration, the United States had contributed $350 million to the Inter-American Development Bank, an agency that would make low-interest loans for economic development projects and had pledged $500 million for a new Social Progress Trust Fund. Yet, the catalyst for change was not solely the perception in the United States that relations with the southern neighbors had soured. Commitments of money came after Washington decided that the upheaval in Cuba, which followed the overthrow of dictator Fulgencio Batista on January 1, 1959, portended grave dangers for the United States. The communist revolution in Cuba led by Fidel Castro, if emulated throughout Latin America, would threaten U.S. financial interests as well as imperil the nation's security.[67]

A fear of communist penetration also motivated the new U.S. policies toward post–Pérez Jiménez Venezuela. The official explanation for the attack on Vice-President Nixon, in which members of the Venezuelan Communist party participated, blamed "a small Communist minority," although Nixon admitted in a speech to the National Press Club that "while it is true that Communists spearheaded the attacks, they had a lot of willing spear carriers with them."[68] But what most worried the Department of State was that Venezuela would be the object of "a quiet, elaborate Communist infiltration." The provisional government allowed all parties, including the Communist, to function freely and openly. Moreover, Admiral Larrazábal, who became a candidate for president on the Unión Republicana Democrática ticket, accepted the aid of the Communists and thereby ignored the blunt warning he received from Nixon in May 1958 that "freedom cannot survive in any coalition with communists."[69] The department therefore concluded that Rómulo Betancourt must win the election. Through former Assistant Secretary of State Adolf Berle, who knew Betancourt well, the department offered the Venezuelan leader aid, including presumably covert assistance. The stakes seemed high, for, as the Central Intelligence Agency told Berle, "the head of the Communist Party has been to Moscow, has been told that the Venezuelan experiment is first on Moscow's American list, and is going to work at it in this fashion [through Larrazábal]." But a confident Betancourt rejected Berle's offer.[70] In the election, he garnered 49 percent of the vote, while Larrazábal won 35 percent, and Rafael Caldera of the Christian Democrats polled only 16 percent. The suspect Larrazábal congratulated Betancourt and urged his disappointed supporters to accept the electoral results. Betancourt also received congratulations from President Eisenhower after Acting Secretary of State Christian Herter urged him to do so, noting that Betancourt's "closest rival for the Presidency has accepted the support of the Communist Party."[71]

The Eisenhower administration's embrace of Betancourt was not simply an extension of its policy of backing any anticommunist government in Latin America. As Berle recorded in his diary, "this is quite a change from the days" when the Department of State "kicked him around." Truly, a special irony attached to Washington's perception of Betancourt as a savior of Venezuela, for during the mid-1950s, the department shunned the exiled politician and once, complying with a whim of Pérez Jiménez, ordered him out of Puerto Rico.[72] But what was particularly significant was the department's decision to favor, in Secretary Herter's words, "the leader of the leftist but anti-Communist Democratic Action Party." Betancourt's election could pose problems for U.S. interests in Venezuela, for, while he pledged to respect existing investments, he also resurrected Pérez Alfonzo's ideas, promising to raise taxes, restrict oil production, establish a national oil company, and ban new concessions.[73] The election of Betancourt would also mean, however, that a party committed to evolutionary social and economic reform would rule. To the hard-pressed State Department, this now seemed preferable to a right-wing government. As late as November 1957, Secretary of State Dulles asserted that the problem of communism in Latin America was not "in any degree alarming" and that he saw "no likelihood at the present time of communism getting into control of the political institutions of any of the American Republics."[74] A year later, Washington was alarmed and a consensus was emerging, which the fall of

Batista and the rise of Castro seemed to support, that conservative, anticommunist governments, by frustrating legitimate aspirations for social justice, forced desperate people to turn to communism. Support for the deposed tyrants, like Gustavo Rojas Pinilla of Colombia, Manuel Odría of Peru, Batista, and Pérez Jiménez might have been inimical to U.S. interests in Latin America.[75]

By the end of its second term, the Eisenhower administration had laid the foundation for the Alliance for Progress built by President John F. Kennedy. The United States now proclaimed its preference for representative governments that pursued social and economic reforms. As President Eisenhower, shortly before leaving office, told the new Venezuelan ambassador to the United States, "I am convinced that the just aspirations of our fellow citizens in Latin America to improve their social and economic lot can be achieved without the sacrifice of the free and democratic institutions which form the basis of our Western civilization." Moreover, the United States stood ready to encourage and assist these governments; public aid would be available to supplement private investment. These measures would help defeat the "extremists of both right and left."[76] What remained unmodified in the Latin American policy of the United States, however, were the assumptions that no fundamental differences existed between the interests of private U.S. investors and their Latin American hosts and that the prices raw-material producers received from their industrial customers must be determined by the free play of the world economy. Venezuelan challenges to these traditional U.S. policies would dominate relations between the two countries during the 1960s.

7. The Alliance for Progress, 1959–1968

Between 1959 and 1968, the United States saw relations with Venezuela as critically significant. Certain that the Latin American nation had become a Cold War battleground, the United States rushed economic and military assistance and diplomatic support there. Moreover, it again pointed to its diplomacy toward Venezuela as a model of its Latin American policies; aid was given to a government committed to the goals of the Alliance for Progress, economic and social reform within a constitutional framework. Yet, though the two countries worked together through the 1960s, tension increasingly characterized their relationship. Venezuela not only resumed its campaign to regulate the foreign oil industry but also began to challenge the foreign economic policies of the United States. To the Venezuelans, liberal trade practices meant that the prices they received for their raw-material exports continually declined, thereby retarding economic growth and undermining the alliance program.

With the inauguration in February 1959 of Rómulo Betancourt as president, Venezuelans tried political democracy for the second time in the twentieth century. Their new leader was a veteran of the first attempt. President Betancourt was born in 1908 in the small town of Guatire in the coastal state of Miranda. His parents lived modestly. Betancourt's father was an immigrant from the Canary Islands while his mother was a native Venezuelan of mixed racial origins. After attending the Liceo Caracas, where he studied under novelist Rómulo Gallegos, he matriculated at the Universidad Central de Caracas. But he never attained his law degree, for as a leader of the "Generation of '28" he had to seek exile after the suppression of the student protests against the Gómez tyranny. After another brief exile during the López Contreras regime, Betancourt helped organize the Acción Democrática party. His first opportunity to wield political power came between 1945 and early 1948 when he headed the junta during the *trienio*. Following the overthrow of the Gallegos administration in November 1948, Betancourt fled the country and lived in Costa Rica, Cuba, Puerto Rico, and New York. While in exile, he kept in contact with Acción Democrática's clandestine organization in Venezuela. He also exchanged ideas and plans with other members of Latin America's "democratic left," such as Juan Bosch of the Dominican Republic and José Figueres of Costa Rica. In addition, he wrote his fiery *Venezuela: Política y petróleo*, a scathing indictment of Pérez Jiménez and the foreign oil industry.[1] His triumphal return to Venezuela and resounding electoral victory in 1958 were gratifying personal victories for a politician who had spent twenty of his fifty years in exile.

As they were during the *trienio*, the goals of Betancourt and his ruling AD party were economic diversification, agrarian reform, improvements in educational and social welfare, and the consolidation of constitutional democracy. While the ends of public policy remained the same, the means that President Betancourt and his advisors chose had changed; they now favored a moderate, evolutionary approach to economic and social reform. The *adecos* contemplated neither fundamental tax reform nor any significant redistribution of wealth. Betancourt's four-year plan was a form of state capitalism; public money would be invested by a central authority in state as well as private industrial enterprises. Agrarian reform would be similarly accomplished. The agrarian reform law of 1948 encouraged local peasant organizations to determine the pattern and pace of land redistribution. The law of 1960 created a centrally controlled, planned process that would parcel out land only as support services, such as roads, homes, and marketing facilities, were in place. The state would compensate the original landowners, and it would also provide subsidies to commercial farmers.[2]

Betancourt and other senior *adecos* pursued pragmatic policies based on their post-1948 analysis of the Venezuelan political milieu. Ten years of military dictatorship had been a harsh and bitter experience for both them and the country; the party had been outlawed, leaders imprisoned and exiled, and some members butchered by the military. The chastened *adecos* were now reluctant to alienate any powerful interest, such as estate owners, business groups, or the church. Using oil income, they hoped to work around or through the country's powerholders. For example, the AD governments would grant more financial aid to the Catholic church than had Pérez Jiménez. The *adecos* feared that an aggrieved group might appeal to the military to intervene again. Keeping the armed forces in the barracks meant that civilian unity must be maintained. Before the December

1958 election, the leaders of AD had agreed in the Pact of Punto Fijo with the other two major parties, the Christian Democrats (COPEI) and Unión Republicana Democrática (URD), that Venezuela would be ruled by a new coalitional government under the leadership of whichever party won the presidency.[3] In particular, the victorious *adecos* needed the assistance of COPEI, a Catholic party that drew on the ideals of papal encyclicals, such as *Rerum Novarum* of Leo XIII, entreating the state to promote both the social and the spiritual welfare of the citizen. COPEI's role in the government lent a further leavening of moderation to AD's policies.[4]

As the *adecos* surmised, the Betancourt government faced during its first three years in office a series of threats from right-wing groups. Disgruntled military and civilian *perezjimenistas* hatched numerous conspiracies. In June 1960, President Betancourt was badly burned and nearly assassinated when, as his limousine passed a parked automobile, a bomb implanted in the automobile was exploded by a remote-control device. Exiled military officers organized the plot with the assistance of Rafael Trujillo, dictator of the Dominican Republic and erstwhile ally of Pérez Jiménez. This assassination attempt and other subversive activities failed because Betancourt held the support of influential civilian groups and most military officers. He constantly assured the armed forces that he was sympathetic to their institutional needs, he approved liberal defense budgets, and he appointed high-ranking officers to ambassadorial and governmental posts.[5]

Even as it struggled to prevent another conservative military *golpe*, the Betancourt government encountered new threats from the political left. Violent antigovernment demonstrations rocked Caracas between October and December 1960. These riots were fomented by youthful members of AD and the URD and Venezuelan communists, who were frustrated by the pace and tenor of change in the nation. These young Venezuelans, who had not fled the country during the 1950s, had become politically radical while resisting the military regimes. After surrendering leadership in 1958 to the old guard of their parties, they were shocked by their elders' political moderation. They charged that the Betancourt government, in placating elites as well as the military, had abandoned meaningful reform, such as the rapid redistribution of land. For the disenchanted politicians, many of whom joined the Movimiento de Izquierda Revolucionaria (Movement of the Revolutionary Left, or MIR), the Cuban Revolution was the answer.[6] As one spokesman emphasized, "we saw that while our leaders had been talking about revolution for thirty years, in Cuba the Revolution triumphed in two years of fighting."[7] The disorders of late 1960 ushered in nearly a decade of leftist agitation which included guerilla warfare and urban terrorism.

The United States intervened to save the beleaguered Betancourt government. At a meeting of the Organization of American States held in San José, Costa Rica, in August 1960, the United States joined with eighteen other American nations in breaking relations with the Dominican Republic. Outrage over Trujillo's vicious attack on President Betancourt overcame what disinclination the Eisenhower administration had about jilting a dictator who had fawned U.S. power for thirty years. In addition, the U.S. delegation hoped that the Dominican resolution might serve as a precedent for isolating Castro's Cuba. When Venezuela charged that Trujillo continued to conspire against it, the new presidential administration of John F. Kennedy imposed in early 1961 stiff economic sanctions against Santo Domingo. That crisis culminated in May when Dominicans assassinated Trujillo.[8] But what particularly alarmed Washington were the leftist-inspired riots of late 1960 in Caracas. Foreign-policy advisors to the president-elect frantically prepared for what they feared might be the second communist beachhead in the Americas.

On March 13, 1961, less than two months after his inauguration, President Kennedy unveiled the Alliance for Progress, the new U.S. approach to Latin America. In that speech and in a subsequent inter-American planning meeting held in August at Punta del Este, Uruguay, the United States announced that it was prepared to transfer as much as $20 billion public and private money to Latin America within ten years. This infusion of capital combined with technical aid would help each nation achieve a real economic growth rate of at least 2.5 percent per capita a year. In return, Latin American delegates pledged to take the hard measures to ensure that the new prosperity benefited all; they promised to reform tax codes, redistribute income and land, expand educational opportunities, and improve health services and to do so by democratic processes. Compassion for Latin America's needy millions, a desire to repair inter-American relations, and a concern for national security motivated the new U.S. program for Latin America. The ambitious alliance would be, in Secretary of the Treasury C. Douglas Dillon's words, a "controlled revolution," providing a democratic, capitalist alternative to Cuba's revolutionary socialist formula for economic development in Latin America.[9]

Analyses of Venezuela's turbulent political scene significantly influenced the formulation of the alliance. Following his election, Kennedy, who had during the campaign sharply criticized Eisenhower's Latin American policy by citing the Nixon incident in Caracas and the rise of Castro, commissioned a task force to prepare policy recommendations.[10] The study group, headed by Adolf A. Berle, warned the president-elect that he must act quickly and boldly, for "the Cold War is on in Latin America" and that it could be lost "if the United States is a spectator." The "point of greatest attack" was Venezuela; the Central Intelligence Agency (CIA) informed the task force that Venezuela was a "hinge to the situation" and that a successful communist uprising in Venezuela might fan "brushfire wars" all along the Caribbean littoral. The task force called for emergency measures, including the "spectacular assignment" of a new ambassador for Venezuela, economic aid, and military support. As the task force's memorandum to Kennedy concluded, "we are in a political fight and cannot run away from it by hiding behind the doctrine of non-intervention."[11]

Fortunately for the United States, Venezuela was led by Rómulo Betancourt, a politician presidential advisor Arthur Schlesinger described, after touring South America in February 1961, as "the most impressive" in Latin America and one both he and Berle approvingly dubbed as a "New Dealer." As the chief of a democratic political party, Betancourt would also be, in the opinion of the task force on Latin America, an effective "political instrument with which to fight the 'Cold War' on the streets, outside the limitations of formal diplomacy." Moreover, the Venezuelan was a strident anticommunist. In his book, he denounced international communism, which he considered incompatible with Venezuelan nationalism, with nearly as much fervor as he attacked Pérez Jiménez. As president, he refused to include the Venezuelan communists in the ruling civilian coalition, declined to establish relations with the Soviet Union, labeled the urban riots in late 1960 as "Communist-inspired," and was guiding his nation toward an open break with Cuba. With its argument that "we need Rómulo Betancourt to keep things in line," the task force dismissed the theory, which prevailed during the Dulles years, that authoritarian governments could best combat communism in Latin America.[12]

The task force's report, which Kennedy received in January 1961, served as a basis for the Alliance for Progress. Yet, the United States needed a bolder approach than merely proclaiming its support for progressive, democratic parties, if it wanted to compete for the allegiance of Latin Americans. The task force, which was reconstituted in February 1961 under the direction of Berle and Department of State officer Thomas Mann, worked on a Marshall Plan for Latin America. The group solicited advice from prominent Latin Americans, including José Antonio Mayobre, Venezuela's ambassador to the United States, and Betancourt.[13] Kennedy's advisors not only listened to Venezuelans in developing a plan to underwrite economic development and social change in Latin America but also saw Venezuela "as a model for Latin American progressive democracy." What the proponents of the alliance wanted was to "modernize" Latin American society through a "middle-class revolution." This meant that the agrarian, semifeudal economic structures, which primarily benefited the landholding oligarchy, must be eradicated. According to prominent U.S. social scientists, the place of the oligarchs would be taken by the burgeoning new urban middle classes, now perhaps 25 percent of Latin America's population. Once in power, these middle-class revolutionaries, such as the *adecos*, would foster industrialization and economic growth and the concomitant features of a modern technical society, such as constitutional government, bureaucratic efficiency and honesty, and social mobility. As presidential advisor Walt W. Rostow, a professor of economics and history and author of the influential *The Stages of Economic Growth*, told Kennedy, U.S. economic assistance would help underdeveloped nations, such as Venezuela, move beyond the "take-off" stage into self-sustaining economic growth in the 1960s. Indeed, as Schlesinger warned the president, the United States must promote the middle-class revolution, for "*if the possessing classes of Latin America make the middle-class revolution impossible, they will make a 'workers and peasants' revolution inevitable*; that is, if they destroy a Betancourt, they will guarantee a Castro or a Perón."[14]

Under the aegis of the Alliance for Progress, the United States and Venezuela cooperated closely as symbolized by the installation in early 1961 of a direct telephone line between the White House and Miraflores. Presidents Kennedy and Betancourt frequently discussed both bilateral and inter-American relations and a mutual admiration quickly developed. Betancourt congratulated Kennedy on his alliance speech, declaring that he spoke "a language that has not been heard since the days of Franklin Delano Roosevelt."[15] Venezuela was the first Latin American country the young leader of the United States visited. During his tour in December 1961, Kennedy was greeted by large, friendly crowds, and he participated in a land redistribution ceremony in the rural town of La Morita. Betancourt returned the courtesy when he traveled to Washington in February 1963. There he heard himself praised by Kennedy as one who represented "all that we admire in a political leader" and his accomplishments described as "a symbol of what we wish for our own country and for our sister republics."[16]

Kennedy and his successor, Lyndon B. Johnson, supplemented

this verbal support with significant political, economic, military, and covert assistance for the Acción Democrática governments. Through the 1960s, the United States persistently attempted to impress all shades of the Venezuelan political spectrum that it was committed to the preservation of representative government and social and economic progress in Venezuela. Washington's representatives in Venezuela, led by Ambassador C. Allan Stewart, continuously emphasized to influential Venezuelans that the United States favored the Betancourt government.[17] President Kennedy similarly lobbied Venezuelan visitors to Washington. He praised Rafael Caldera, leader of the Christian Democrats, for his party's contribution "to political stability and the continuing process of democratic evolution in Venezuela," and he also entertained Defense Minister Brigadier General Antonio Briceño Linares and pointedly told him, while conferring about the general's shopping list of military equipment, that "he was aware of the support given by the Armed Forces to President Betancourt." After the suppression of a serious military uprising in mid-1962, Kennedy publicly congratulated Venezuela. His press release, written by Stewart and approved by Betancourt, was calculated to reiterate U.S. "opposition to military coups against constitutional processes."[18] The Johnson administration also heartily supported a "Government which is in the forefront of political and economic democracy in Latin America and the Alliance for Progress effort."[19]

To help Venezuelan democracy survive, the United States also designed, under the Alliance for Progress, an extensive economic aid program for the South American nation. The political agitation that President Betancourt faced was exacerbated by a deep economic recession which gripped Venezuela during the early 1960s. Unemployment rose officially to 15 percent, although it was much higher in the ever-expanding *ranchos* of Caracas and Maracaibo. These miserable shantytowns were fertile grounds for organizers from the political left. To help Betancourt calm urban areas, the Kennedy administration, as the task force recommended, rushed an emergency package of $100 million of aid to Venezuela in early 1961. Between 1962 and 1965, U.S. agencies, such as the Agency for International Development (AID), granted or loaned over $140 million to Venezuela. Venezuela obtained another $200 million in credits, with the backing of the United States, from international lending agencies, particularly the World Bank and the Inter-American Development Bank. The government also secured loans from private U.S. bankers. With the money, it attacked urban problems by initiating public housing and public works projects.[20]

Beyond generating public money for Venezuela, the Department of State also encouraged private companies to invest there. Defining one of the prime objectives of the alliance as the "encouragement of private investment in helping build [the] economic infrastructure [of] Latin American countries," department officials met with businessmen in New York and urged them to enter Venezuela. They conceded that political turbulence in the nation made investments seem risky but that their caution, while understandable, "was playing Castro's game." Moreover, if foreign businessmen displayed confidence in Venezuela, it would inspire native entrepreneurs to invest at home.[21] Officials, such as William Gaud, the administrator of AID, also wanted capitalists to invest in areas other than extractive industries in order to help the host nation develop enterprises that would be labor-intensive, stimulate other sectors of the economy, develop internal markets, and help free it from the vicissitudes of world trade.[22] During the 1960s Venezuela received some help in diversifying its economy. U.S. investments in manufacturing, for example, rose from $180 million to $500 million. Most investments, however, continued to be in the extractive industries of oil and iron ore development.[23]

While U.S. officials predicted that the Betancourt government could not survive unless it ameliorated social and economic problems, they also believed that the armed forces would be "the most important single element in determining whether Betancourt will last until 1964 and be succeeded by another freely elected chief executive for [the] first time in Venezuelan history."[24] Accordingly, the United States helped President Betancourt appease the military. Between 1961 and 1965, the United States supplied over $60 million in credits and grants for military equipment and training, twice the amount of military aid supplied during the 1950s. It trained Venezuelan officers at the Special Warfare Center at Fort Bragg, North Carolina, where the heralded "Green Berets" were stationed, and it trained over one thousand troops at the Jungle Warfare School in the Canal Zone.[25] High-ranking officers were invited to Washington to attend military colleges, like the Inter-American Defense College, in the expectation, in President Kennedy's words, "that close association with American military, who understood so well the need to subordinate the military power to the civilian, would be helpful" in depressing aspirations for political power.[26]

What the United States refused to do in assisting Betancourt control the armed forces was to lend unqualified support to his policy of recognizing only "regimes born of free election and respecting human rights," the Betancourt Doctrine. The policy, which resembled the Larreta Doctrine of 1945, was founded on the premise that it was "nonsensical" to denounce totalitarian regimes in Asia and Europe and tolerate despotic governments in the hemisphere, as in 1954 when the American states condemned international communism while meeting in Pérez Jiménez' Venezuela. Betancourt also hoped the doctrine would notify the Venezuelan military that, if they struck another *golpe,* they would encounter united opposition from all American republics, including the United States. During the 1960s, Venezuela, under Betancourt and his successor, Raúl Leoni, severed relations with nine Latin American countries.[27] Although committed by the Alliance for Progress to promote democracy, the Kennedy administration declined to follow the Venezuelan lead; it acquiesced in military *golpes* in Argentina, Ecuador, and Guatemala while objecting to those in Peru, Honduras, and the Dominican Republic. It also rejected a Venezuelan proposal that a special meeting of the Organization of American States be convened in 1962 to reaffirm "democratic solidarity." The reaction of the United States seemed to depend on how the military action affected private U.S. interests and, in particular, whether it promoted or stunted the growth of communism. For example, Secretary of State Dean Rusk apologized to Betancourt for the U.S. decision to recognize the new Argentine government after the fall of Arturo Frondizi but then noted that "my fear is that the failure to do this will only play into the hands of the extremists."[28] During the Johnson administration, the United States, under the leadership of Assistant Secretary of State Thomas Mann, virtually ignored the Betancourt Doctrine and in 1964 probably even encouraged the Brazilian military to overthrow the civilian president, João Goulart.[29]

The ambivalent attitude that the United States demonstrated toward the armed forces indicated that it considered military influence in Latin American politics more than just an unfortunate "fact of life." U.S. aid was intended not only to help President Betancourt but also to upgrade the Venezuelan armed forces' fighting abilities. Transfers of light arms, pistols, submachine guns, and grenades assisted the military in internal security and counterinsurgency operations against the political left. Moreover, the U.S. military mission in Venezuela, which was the largest in Latin America, trained men in antiguerilla tactics and, as in South Vietnam, assigned personnel to advise in combat operations against guerillas.[30]

The United States supplemented its military assistance with other internal security programs. Both the CIA and the AID trained Venezuelan police on how to resist urban terrorism and exchanged information with Venezuelan officials on suspected subversives.[31] To dissuade potential recruits to antigovernment groups, U.S. officials also conducted, what Adolf Berle tagged, "a psychological offensive" in Venezuela. The United States Information Agency supplied over two thousand hours of programming to eighty of Venezuela's ninety-two radio stations. Newspapers were similarly deluged with the Information Agency's press releases, and the major television networks, which U.S. television networks (ABC, CBS, NBC) partially owned, mainly played programs produced in the United States. In addition, the Information Agency sent books to libraries and bookstores, and it may have funneled financial assistance to university professors considered "pro-American" to help them prepare textbooks for undergraduates and high-school students.[32]

To ensure that all U.S. personnel in Venezuela, ranging from Green Berets to Peace Corps volunteers, cooperated, the Department of State sent an interdepartmental team to Venezuela in September 1962. The team, which was modeled after the Taylor-Rostow mission to South Vietnam in 1961, included senior representatives from the Departments of Defense and Justice, Information Agency, AID, and CIA under the leadership of former Ambassador to Afghanistan Henry Byroade.[33] The composition of the team and the disparate projects they investigated—irrigation systems, steel mills, military installations, schools—suggested that the United States was intent not only on saving President Betancourt but also in building a modern nation. The middle-class revolution would be hastened, or, as Walt Rostow put it, "modern societies must be built and we are prepared to build them."[34] When finished, as one AID official wrote to presidential advisor Schlesinger, "the success of Venezuela under a democratic government might well be compared to Cuba's worsening economic position and deteriorating social structure under a Communist regime."[35]

The infusion of people and programs from the United States came as little surprise to the Venezuelan government. Except perhaps for covert activity, President Betancourt and high-level political and military officials knew and approved of U.S. policies and actually initiated some, as when the National Guard requested the training services of a guerilla warfare team.[36] Betancourt proved especially adept at winning aid for his nation. Like Pérez Jiménez, he knew how to touch Washington's most sensitive nerve. When he wanted more money, he could either tease Kennedy about those seeking to "establish in Caracas a branch office of Habana" or blast the AID for not quickly funding a slum clearance and community development project that "would provide the nation with stable bases and, incidentally, would protect it against subversion by reactionaries and pseudo-leftists." Nor was he remiss in reminding the U.S. press and public that Venezuela stood with the United States in "common dedication to the Western democracies and Christian values."[37]

Betancourt's shrewdness also sustained him at home as he picked his way carefully between political extremes. The turmoil generated by the political left ironically may have benefited the president, for, while the Cuban Revolution inspired some Venezuelans, it terrified others. Betancourt artfully reminded military officers that Fidel Castro destroyed the Cuban army and replaced it with a people's militia. The exodus of the Cuban upper classes to Miami also helped convince Venezuelan businessmen that Betancourt's brand of moderate reform might be the best way to protect their property. Church leaders similarly decided to work with a government increasingly seen less as a socialist menace and more as a bulwark against atheism and communism.[38]

President Betancourt and his fellow *adecos* not only relied on analogy to woo political conservatives but also mollified them by repressing radicals. They purged members of the MIR and communists from the peasant and oilworkers unions, supplied the oil companies and other businesses with lists of extremists they should fire, banned leftist student organizations in public secondary schools, and dismissed leftist teachers in primary and secondary schools. When the radicals accelerated the violence, which included in 1962 the killing or wounding of eighty policemen, robberies, burning of buses, sabotage in the oilfields, and collusion in two bloody uprisings with leftist-oriented marines at Carúpano and Puerto Cabello, the government suspended the parties and arrested their leaders. In 1963, when the radicals assaulted military personnel, the government responded to military demands by authorizing mass arrests of leftists, revoking the congressional immunity of senators and deputies from the MIR and the Communist party and turning them over to military courts, and permitting the armed forces to combat the radicals in the streets. In his campaign against the political left, President Betancourt suspended constitutional guarantees five times, 778 out of the 1,847 days he was in office.[39]

To political moderates, *adecos*, and Christian Democrats, the price Betancourt paid for his survival seemed justified when on December 1, 1963, over 90 percent of the electorate came to the polls despite the threat by leftists to shoot anyone who voted. Raúl Leoni, an *adeco* and member of the "Generation of '28," captured a plurality of the vote and was inaugurated president on March 11, 1964. Betancourt's gift of the presidential sash to Leoni was a watershed in

Venezuelan political history; with a popularly elected president completing his term for the first time, it set the nation on a course of orderly transfers of power.

Venezuela's national election earned applause from the United States. President Johnson entertained Betancourt in Washington shortly after the Venezuelan retired to "demonstrate our recognition of the importance of the Venezuelan triumph of democracy over communism and our support of that country's enlightened leadership which has put it in the vanguard of the Alliance for Progress."[40] Despite Washington's self-congratulatory tone, that an elected government lasted in Venezuela probably depended more on the policies pursued by a skillful politician than on the amount of assistance the United States provided. While confident of their ability to build a nation, President Kennedy, the director of the CIA, and State Department officials soberly concluded between 1961 and 1963 that "political stability seems to depend on one heart, that of President Betancourt."[41] Whether Betancourt's decision to preserve the constitution by placating conservative elites at the expense of far-reaching reform for the majority of Venezuelans, peasants and *rancheros*, was wise is moot.[42] Based on the *trienio* experience, the hard years of exile, and the Pact of Punto Fijo, the leaders of Acción Democrática seemed inclined to follow that path, before hearing of the promises of the Alliance for Progress. The United States, of course, designed the alliance with the Venezuelan party in mind. Yet, U.S. support would have been circumscribed had the Betancourt administration accommodated the radicals.

While carefully refraining from directly meddling in Venezuelan politics, U.S. officials made Venezuelans aware of their ideas. In May 1961, for example, President Betancourt announced that he would fight the deep economic recession with severe budget cuts, new taxes, and reductions in bureaucrats' salaries. The austerity measures, which further alienated the MIR and communists, came as the government was negotiating loans from both the United States and the World Bank and was hearing about the virtues of fiscal responsibility from U.S. and international officials.[43] Such budgetary restraints allowed Ambassador to Venezuela Stewart to assure skeptical U.S. congressmen that, while Betancourt once "had a definite Socialist type of thinking," he now ran a "middle-of-the-road government. Absolutely." Stewart also advised the government on public relations, telling Minister of Interior Carlos Andrés Pérez in mid-1962 that arrests of radicals "needed dramatizing." As he wired to Washington, "I urged him to make this known to people of country in radio-television broadcast, saying it was my impression most Venezuelans did not know GOV [Government of Venezuela] was taking energetic steps."[44]

Where the United States particularly exercised influence in Venezuela was in its relationship with the armed forces. The United States had other interests beyond helping the government court the military and training and equipping soldiers for counterinsurgency operations. While frequently lamenting the role that military officers played in Latin American politics, the Kennedy administration was committed to preserving the military's power. As Ambassador Stewart recorded in reviewing the Military Assistance Program: "We must recall that in the long run US-oriented and anti-Communist armed forces are vital instruments to maintain our security interests in Caribbean region. Probably most important single difference between Cuba, where Communist dictatorship has taken over, and Guatemala (1954), Venezuela (1958–1962), and the Dominican Republic (1961), where there have developed relatively democratic systems, is that the Cuban armed forces disintegrated while the US-oriented armed forces of the others remained intact and able to defend themselves and theirs from Communists. Had Cuban armed forces done same, large scale 1962 [sic] invasion from outside would probably have been unnecessary." In addition, Stewart, citing the revolts by marines at Carúpano and Puerto Cabello in mid-1962, reminded the department that "it appears highly dangerous to assume that in absence actual or anticipated US support, Latin American military so right wing it will not affiliate with local Communists."[45] Indeed, as a scholar has argued, the Kennedy administration lashed out at the annulment of elections in July 1962 by the Peruvian military not just because one of Washington's favorite middle-class revolutionaries, Víctor Haya de la Torre, had been denied victory, but also because officials judged Peruvian officers ultranationalistic and soft on communism.[46]

What the United States displayed in its policies toward the Betancourt government was one aspect of the ambivalences and uncertainties that plagued the Alliance for Progress and ultimately condemned it, critics have charged, to failure. The task force on Latin America recommended that U.S. diplomacy be based on "clear, consistent, moral, democratic principles" and that the nation be prepared to tolerate "agitations and disturbances" that would surely follow as the privileges of elites were challenged. But, in the end, the program's anticommunist bent, what Arthur Schlesinger called its "edge of toughness," became the overriding consideration.[47] Aggrieved powerholders linked social instability with communism and won Washington's sympathy. As Chester Bowles, a disgruntled member of the Kennedy administration, observed, the United States needed to encourage and identify with social revolution in Latin America. But "Jack Kennedy was not up to that. He might have come up to it, but you couldn't patch up this and patch up that and please Rusk and please Standard Oil and please somebody else all at the same time." As a result, "we never quite decided whether we were prepared to put up with dictators in a pinch, or whether we were not prepared to put up with them at all."[48] As such, the United States adopted the view propounded by Thomas Mann even during the deliberations of the task force "that Latin American armies constituted the only real impediment to Communist take-over in most of Latin America."[49] The civilian governments of Venezuela proved an exception to Mann's rule, and the United States stuck with them.

The failure of extremists to intimidate Venezuelan voters and the inauguration of Leoni eased U.S. fears about Venezuela's future. By the mid-1960s, the United States was dismantling its extensive economic assistance program in Venezuela. Between 1966 and 1969, AID assistance, for example, amounted to only $5.5 million, $42 million less than in the previous four years. With their economy beginning to show moderate rates of growth, Venezuelans had, in President Johnson's words, "progressed to the point where they no longer needed AID loans."[50] While Venezuelan officials did not share Johnson's optimism, they could not, in any case, obtain sufficiently attractive loans from the United States. Alarmed by mounting balance of payment deficits and unbalanced domestic budgets, Congress raised interest rates, shortened amortization periods, and attached other conditions to international loans. Like the Great Society, the Alliance for Progress became a casualty of the war in Indochina. The effect, according to Ambassador Maurice Bernbaum, who was assigned to Caracas in 1965, was to reduce the sense of U.S. power and presence in Venezuela.[51]

While confident that the Leoni government could maintain stable, orderly rule, the Johnson administration continued to assist it in fighting guerillas and choking off their sources of supply.[52] On November 3, 1963, less than a month before national elections, Venezuela announced that it had found a cache of arms, three tons of rifles, machine guns, mortars, and bazookas, on a secluded beach in the state of Falcón. The Foreign Ministry subsequently charged, based on serial markings on the weapons and intercepted messages, that Cuba had left the arms on the beach for terrorists. Upon a formal complaint from Venezuela, the Organization of American States in July 1964 recommended to its members, by a vote of 15 to

4, to break relations with Cuba and impose economic sanctions on it. Three years later, the organization again condemned Cuba after Venezuelan officials captured on their shores four Cuban military officers who had sailed there from Cuba with a group of Venezuelan guerillas. In each case, the United States vigorously lobbied other American states to isolate Cuba from the inter-American community.[53]

Years of hostility lay behind the Cuban intervention in Venezuelan affairs and the Venezuelan diplomatic offensive against Cuba. As a revolutionary, Fidel Castro seemed determined to demonstrate that the reform-oriented Acción Democrática governments, the pride of the Alliance for Progress, could not improve the lives of the poor. In addition, he felt a special responsibility to the young Venezuelan radicals who looked to him for guidance. From Havana flowed a steady stream of invective: encouragement to the guerillas and terrorists and unseemly personal attacks on President Betancourt.[54] Yet, Castro's diatribes were not entirely unprovoked. While Cuba was training Venezuelan insurgents, Cuban exiles were probably plotting in Caracas. Moreover, like Castro, President Betancourt saw his programs as proper for all of Latin America. The Venezuelan repeatedly compared the repression in Cuba to the freedoms of Venezuela. And he had assisted the anti-Castro policies of the United States. He reportedly knew about and approved of the planning for the Bay of Pigs operation, and he assigned two destroyers to the naval blockade of Cuba during the missile crisis.[55]

The United States repaid its anti-Castro ally. Unlike economic aid, military assistance remained available through the 1960s. In 1966, a year in which the government mounted a major offensive against insurgents, Venezuela received nearly $12 million in military grants and credits. The next year, President Johnson, at Leoni's request, "arranged promptly a mission to Venezuela to clarify precisely what you need to deal with your guerilla problem and how we can be of help to you." In addition, Washington thrice ordered reconnaisance planes to scour the Caribbean as a warning to Cuba not to ship arms to Venezuelan guerillas. With this aid, the Ministry of Interior and the armed forces between 1964 and 1967 methodically pursued the urban and rural insurgents. The guerillas, whose strength intelligence agencies estimated in 1965 as less than fifteen hundred, suffered a series of blows from the government and by 1967 had virtually ceased their agitation.[56] The classic guerilla strategy of living with and arousing peasants failed in Venezuela, for by carrying out some land reform the Acción Democrática governments pacified the countryside. In the end, the leftists' only sanctuary was the Universidad Central in Caracas. They lost that when, in 1967, President Leoni breached the traditional autonomy of the university, allowing the armed forces to occupy the campus.[57]

While Cuba violated international law by meddling in Venezuelan affairs, both Washington and Caracas exaggerated the threat by loosely and interchangeably using the terms "Communist," "international communism," and "Castro-Communism." White House officials often put Venezuela within a global context as when Walt Rostow defended military involvement in Vietnam by explaining that "it is on this spot that we have to break the liberation war—Chinese type." Rostow added that "if we don't break it here we shall have to face it again in Thailand, Venezuela, elsewhere."[58] Neither the Chinese nor the Russians evinced much interest, however, in Venezuela. Except for editorial approval in *Pravda*, the Soviets gave virtually no aid to Venezuelan communists, notwithstanding Nikita Khrushchev's 1961 pledge to support wars of national liberation. By mid-1965, the Soviets, apparently attempting to improve their image in Latin America, were counseling communists to cease subversive activities.[59] Soon the Venezuelan Communist party withdrew from the fighting, although the decision was based on the heavy casualties the party suffered at the hands of the government, not directions from Moscow. The new party line was denounced by Castro. Venezuelan communists replied that they rejected "the role of revolutionary 'Pope' which Fidel Castro has arrogated to himself." Castro could therefore only assist members of MIR and a few dissident communists who carried on the fight in 1966–1967.[60] And Cuban aid was always more verbal than material and miniscule compared to the military assistance that Venezuela received from the United States. Moreover, as a CIA study pointed out in 1965, while members of the MIR accepted outside aid, they ran "their own shows," were a "home-grown revolutionary organization," and could be described as an "extreme-nationalist, revolutionary leftist movement." Despite this information, Secretary of State Rusk still claimed, in speaking about Venezuela's plight to the Organization of American States in September 1967, that "Communist dictatorship remains the final Communist objective."[61] To be sure, Venezuelan officials, by constantly portraying their nation as the object of "an aggressive offensive by agents of international communism," encouraged U.S officials to speak forebodingly.[62]

Where Venezuelan-U.S. interests coincided—promoting democracy, containing communism, isolating Cuba—the United States vigorously assisted its Latin American ally. But Venezuela needed more than just AID loans to reach the social and economic goals of the Alliance for Progress, if it relied solely on moderate, evolutionary methods favored by *adeco* leaders and U.S. officials. The alliance was a cooperative venture; it was expected that Latin America would add $80 billion to the $20 billion that the United States contributed to economic development. To uphold the bargain, they would have to draw on export earnings, whether it be from sales of bananas, coffee, copper, or petroleum. The size of these earnings would depend on prices received, profits reaped, and markets opened. In the view of Venezuelans, the U.S. position on the crucial issues of prices, profits, and markets fundamentally undermined their joint aspirations for a "decade of development" in Venezuela.

Within a year after the fall of Pérez Jiménez, the government and the foreign oil industry were again locked in confrontation. On December 20, 1958, shortly after the presidential elections, the provisional government suddenly decreed a revision of corporate income tax laws, thereby raising the oil and steel companies' taxes. Hereafter, the nation would collect at least 60 percent of their profits. While violating the spirit of the oil law of 1943, the government had not broken any legal agreements, since the "50-50" split of profits was not written into the law per se but rather was the result of the exploration, exploitation, income, and royalty taxes that the government collected. Venezuela defended the tax hike by noting that in the previous three years the oil industry had made nearly $2 billion on an investment of $2.55 billion. In truth, the government imposed the tax, which was retroactive for 1958, because it badly needed the extra $176 million. In addition, the provisional rulers may have wanted to present president-elect Betancourt with a satisfying fait accompli. During the campaign Betancourt pledged to expand the nation's oil income, but only after negotiating with the industry.[63]

The decree on December 20, reminiscent of the extraordinary tax levy of December 31, 1945, shocked the oil industry, which had been indulged since the military assumed power in 1948. The industry rebuked the government. The president of Creole Petroleum, Harold "Iron Duke" Haight, charged that the provisional government had broken faith with the companies and that it had a moral obligation to consult with them. Eugene Holman, Chairman of Standard Oil of New Jersey, threatened that "this action will almost certainly call for reexamination of international investments."[64] From Venezuelan officials came the measured reply that "the sovereign government of Venezuela has not yet reached the point where it must obtain permission from Creole Petroleum Company to increase taxes any more than the sovereign government of the United States must obtain permission from Creole's parent, Standard Oil of New Jersey, before income taxes are increased or decreased there."[65] Faced with mounting nationalist fervor, the industry tempered its rhetoric. Standard also recalled Haight and transferred out of corporate headquarters and back to Venezuela the diplomatic Arthur Proudfit, the negotiator of the oil law of 1943.

While the industry stopped publicly challenging the government, it waged through the 1960s a determined campaign to pressure Venezuela into reducing or at least freezing its taxes. Its tactics including retarding the growth of oil production, slowing drilling and exploration, limiting new investments, and cutting back on employment. For example, the number of exploratory wells drilled fell from 598 in 1958 to 75 in 1967, causing the theoretical life of Venezuela's known oil reserves, the lifeblood of the nation, to decline from 17.7 years to 12.3 years.[66] The companies explained that Venezuelan oil was "overtaxed" and that they were satisfying the growing world demand for oil by expanding exploration and production in their more profitable holdings in Canada and the Middle East. As Creole official Leo Lowry once told President Leoni, the government "had the right to enact measures," but "it could not hold the company responsible for the results," for "it was the customer who decided where to buy his oil."[67]

This test of strength between government and industry did not stimulate intervention by the Department of State. Its only direct intercession came at the end of 1958, when, with the memory of the Nixon incident still vivid, it counseled forbearance to oil executives, lest their intemperate remarks force the government to respond with radical measures.[68] In any case, President Betancourt's policies toward the oil companies' activities within Venezuela proved as moderate as his economic reforms. His government agreed that the companies deserved to make a fair profit, which was defined as at least a 15 percent annual return on net fixed assets. In a closed congressional session in mid-1961, his administration defeated a bill to enact new taxes on the oil industry.[69] His only innovations, both campaign pledges, were to establish a national oil company and not to lease new concessions. Neither policy posed any immediate threat to the established oil companies. By waiting until 1960 to found its own oil company, Venezuela had lost nearly forty years of possible experience in the oil business. By the end of the decade, the national company, Corporación Venezolana del Petróleo, produced less than 2 percent of the country's production and its only outlet was the home market. The ban on concessions seemed similarly belated, for Pérez

Jiménez had augmented the companies' forty-year concessions of 1943–1944 with like awards in 1956–1957. The companies scorned the new policies, although, by allowing the nation's proven reserves to contract, they were unsubtly trying to frighten the government into renewing concessions for the rest of the century.

In planning for the distant future when Venezuela might produce its own natural resources, the Betancourt administration was implementing nationalist measures debated between 1936 and 1948. And its retention of the new income tax was an extension of policies pursued by the López Contreras, Medina, and Gallegos governments. Like all other Venezuelan regimes, the Acción Democrática governments conceded, however, that they needed foreign investment and that they could not afford to alienate the companies. As Betancourt recalled, "We were not impractical romantics in Miraflores and in the ministries."[70] What distinguished the oil policies of the Betancourt and Leoni administrations was their determination to ask questions about the fate of their oil once if left the country. The price that their oil commanded and the markets where it was sold increasingly became more important to Venezuelans than the percentage of profits they garnered. Their inquiries directly challenged both the multinational oil companies' conduct of the world oil business and the foreign economic policy of the United States.

For all oil-exporting nations, the fundamental issue of the 1960s was the declining price of a barrel of oil. The average price of Venezuelan crude, for example, fell from a high of $2.65 a barrel in 1957 to $1.81 in 1969. Crude oil actually sold for a lower price than during the 1920s. As explained by the oil executives, the decline in prices was simply a matter of supply and demand. New oilfields in Africa, Canada, and the Middle East proved fantastically rich. Wells in Saudi Arabia, for example, produced 5,000 barrels a day; a good well in Venezuela produced only about 300 barrels a day. Between 1955 and 1965, the combined production of fields in Africa, Canada, and the Middle East rose from 3.6 million barrels a day to a whopping 10.5 million barrels a day.[71]

These rapid changes in world oil supplies rocked the Venezuelan government and economy. Venezuela's international significance had been sharply diminished, for, while it remained the world's single largest exporter, its share of oil exports to the noncommunist world fell from 33 percent in the mid-1950s to less than 20 percent by the end of the 1960s. Moreover, the decline in prices, by reducing the value of royalties, vitiated the effect of raising income taxes. It also disrupted the government's economic development plans as revenue rose only slowly and unevenly through the 1960s. To boost production would be self-defeating, for it would only further depress world prices and jeopardize the dwindling reserves.[72] Finally, these changes seemed to sustain the industry's position that Venezuela was forfeiting its leading role on the world oil stage with its high tax policies.

Even as officials were contemplating the effects of the first round of price cuts, they received another setback when President Eisenhower, using a provision of the Trade Agreements Act of 1955, proclaimed on March 10, 1959, a mandatory quota on oil imports. The executive order was designed to protect the domestic petroleum industry from foreign competition on grounds of national security. Imports of crude oil to states east of the Rocky Mountains would be held to no more than 9 percent of domestic production. While couched in the language of national security, the order was also a response to political pressure. Like Venezuelans, U.S. independent oil producers, who had no foreign holdings, feared that they would be drowned by the abundant, cheaply produced African and Middle Eastern oil about to flood world markets. Venezuela was the nation most immediately affected by Eisenhower's action, however, since it supplied two-thirds of U.S. oil imports. In issuing the order, the White House promised to conduct "informal conversations with Canada and Venezuela looking toward a coordinated approach to the problem of oil." But two months later, Venezuelans were informed that the United States had decided to exempt Canadian and Mexican oil from the quotas on the premise that overland shipments of oil would be secure in wartime while tankers sailing from Venezuela and refineries in the Dutch West Indies would be vulnerable to submarine attacks. It was curious reasoning since the domestic crude the New England and Middle Atlantic states received also came via tankers, which embarked from Galveston, Texas. In any case, under the quota Venezuela's share of the U.S. market fell from 67 percent in 1957 to 42 percent in 1969 (see Appendix, Table D).[73]

Venezuela reacted swiftly to price cuts and import quotas. By 1960, it had instituted a commission to analyze marketing decisions of the foreign oil industry, joined an international agency of petroleum-exporting nations, and launched a diplomatic offensive against the U.S. trade restrictions. As during the *trienio*, Juan Pablo Pérez Alfonzo, the former minister of fomento and now minister of mines and hydrocarbons, directed the new oil policies. The man whose ideas would shape the course of relations between raw-material producers and their industrial customers had an unusual background and tastes. Born in 1903 into a wealthy family, Pérez Alfonzo spent part of his school years in the United States. He later returned to

Venezuela and was briefly imprisoned during the student rebellion of 1928. He eventually attained a law degree and became a successful lawyer and professor of civil law at the Universidad Central. Though he had no training in geology or petroleum engineering, he also became an expert on the oil industry and the unquestioned spokesman within Acción Democrática on oil policies. Pérez Alfonzo's mastery of the technical details of the oil business reflected not only his nationalistic concerns but also his life-style. An ascetic, parsimonious man, he deplored waste of any kind, including the too-rapid depletion of Venezuela's natural resources. He constantly preached the gospel of frugality to his countrymen. By his own rueful admission, he was a Calvinist in a land made spendthrift by the oil bonanza. Like other senior *adecos,* he endured ten years of exile, living for a time in Washington, D.C., after the fall of the Gallegos government.[74]

A decade of reflection had sharpened Pérez Alfonzo's views about the roles of his nation and oil in the world economy. In a time of oversupply and declining prices, he insisted that petroleum, the vital commodity of modern civilization, had an "intrinsic" value above its market price and that it must be husbanded for future generations. His observations of U.S. life during his exile further persuaded him that industrial nations were recklessly wasting a precious resource.[75] Combined with this analysis was his conviction, based on the work of Argentine economist Raúl Prebisch, that Venezuela, like other raw-material exporters, was being victimized by its industrial customers. During the 1950s, Prebisch, who served on the United Nations Economic Commission for Latin America, produced studies that demonstrated that the "terms of trade" moved against raw-material producers; relative to the price of manufactures, the prices of raw materials and foodstuffs had declined. Prebisch's solution, which was an attack on the classical economic theory of trading from comparative advantage, was for Latin American governments to use taxes, tariffs, and subsidies to promote industrialization. While economists debated the causes and consequences of Prebisch's findings and the wisdom of his recommendations, his figures on relative prices for the 1950s were accurate.[76] Until 1958, Venezuela escaped the fate of its neighbors, for, unlike the price of bananas, copper, and sugar, the price of oil kept pace with the price of manufactures. But even owning an ingredient essential to modern industrial societies could not protect Venezuela from a process that Latin Americans passionately believed kept them poor to the benefit of North Americans and Western Europeans.

What Pérez Alfonzo wanted was the power to control prices and production or, in his view, to substitute a public cartel for the private one the multinational oil companies created. At home, this meant the establishment of a monitoring agency, the Coordinating Commission for the Conservation and Commerce of Hydrocarbons, to oversee the companies' production and marketing plans. The commission investigated whether the companies obtained the highest possible price for Venezuelan crude. The commission, ironically, was patterned after the Texas Railroad Commission, a regulatory agency empowered to control oil production to ensure stable prices and generous taxes for the state. The Railroad Commission had gained control over production in the 1930s when desperate oilmen were wastefully selling the state's vital resource for fifty cents a barrel. Venezuelans believed that similarly ruinous competition portended for them. Their agency enjoyed, however, only limited success in the 1960s; the task of compiling independently the data necessary to oversee the industry was a staggering one for a government whose assets were less than the companies, such as Standard Oil of New Jersey and Royal Dutch–Shell, it proposed to regulate.[77]

For a production control or prorationing scheme to succeed, an entity needed market influence, as that exercised in the United States by the Texas Railroad Commission, and political muscle to limit competition, as when U.S. independent oil producers lobbied for import quotas. These lessons Pérez Alfonzo applied to his second innovation, the Organization of Petroleum Exporting Countries (OPEC). On September 10, 1960, Venezuela joined with Saudi Arabia, Iran, Iraq, and Kuwait in a pledge to "study and formulate a system to insure the stabilization of prices by, among other means, the regulation of production." After a sharp cut in August 1960 in the price of crude, Middle Eastern leaders, particularly Saudi Arabia's Oil Minister Sheik Abdullah Tariki, were more receptive to Pérez Alfonzo's entreaties than when Venezuelan officials had toured the Middle East in 1949. They now promised not to allow the multinational companies to use one nation against another. With control of over 80 percent of the petroleum sold on the world market, OPEC would prevent, in Tariki's words, the "dumping or price wars which could result in disaster for the exporting countries."[78]

Despite its potential, OPEC proved weak and ineffective through the 1960s. Rivalries in the Arab world, particularly Iraq's claim of sovereignty over Kuwait, split the organization. OPEC was further weakened by the loss in 1962 of its most vigorous Arab spokesman, Sheik Tariki, who lost favor with the Saudi royal family. Another critical shortcoming was OPEC's inability to restrain new oil producers, like Libya, from cashing in on their bonanza by rapidly expanding production. Criticism and pressure from Western European powers, OPEC's major customers, also restrained the cartel. But the multinational oil companies exposed the fundamental weakness of OPEC by demonstrating that Middle Eastern members, despite pledges of solidarity, had not truly accepted Pérez Alfonzo's ideas. Saudi Arabia and Iran were willing to expand production and accept new price cuts when in late 1964 the companies agreed to give them a slightly higher percentage of profits than the "50-50" split written into their contracts.[79]

The United States scarcely noted OPEC. During the 1960s, the Department of State never felt compelled to issue a major statement on the organization. Indeed, the department, responding to budgetary constraints, actually abolished its petroleum affairs section. This lack of concern was based both on the confidence that in an emergency U.S. oilfields could expand domestic production by over 25 percent and on predictions that great reservoirs of oil within the continental shelf and Alaska awaited exploitation. With the world's chief consumer of oil also its leading producer, it was unrealistic for OPEC to expect to fix prices.[80] U.S. officials and oilmen considered OPEC not only impotent but also irrelevant. With the world awash with oil, the United States ignored Venezuelan suggestions that OPEC and consuming nations work together to control production and increase prices. And oilmen ridiculed Pérez Alfonzo, a man considered friendly toward the United States, for warning that "this manner of free competition will in the long run cause greater harm to consumers since it exhausts an irreplaceable resource more rapidly than necessary, and at the same time limits incentive to search for new reservoirs."[81]

While disregarding OPEC, the United States responded directly to the ideas embodied in the cartel by negotiating with Pérez Alfonzo over his third oil policy initiative, the request for a special preference for Venezuelan oil in the U.S. market. After President Eisenhower announced in May 1959 that Canadian and Mexican oil would be exempted from import quotas, the minister flew to Washington to plead his country's case. In view of his own beliefs in conservation, he approved of protecting U.S. producers from unfair competition. But Venezuela needed protection also. Citing boosts in production during World War II, the Korean War, and the Suez Canal crisis as proof of Venezuela's reliability and friendship, he asked for special treatment.[82] In particular, he wanted the United States, Canada, and Venezuela to coordinate their production and needs. The United States would determine its import requirements and contract directly with the supplying governments. As such, the power of the foreign oil companies in Venezuela to set production and prices would be sharply curtailed; prices would be subject to negotiation between governments. Intergovernmental agreements, by giving Venezuela a dependable market, would also facilitate the growth of a national oil company.[83] While acceptance of Pérez Alfonzo's proposal would fulfill Venezuela's nationalist aspirations, it would also give the United States a dependable and secure oil supply. As Pérez Alfonzo told a congressman from New England, Venezuelans "are waiting for the day when the United States will come back to them."[84]

The United States reviewed the Venezuelan request in 1959, 1962, and 1965, and each time granted the nation concessions while rejecting Pérez Alfonzo's belief that governments, not private interests, should make oil policies. The decision of 1959 defined U.S. policy for the 1960s. After thorough cabinet-level discussions, the United States relaxed slightly quotas on residual fuel oil, a petroleum product used for burning by power plants and factories. The quota on residual oil was susbsequently expanded in early 1963 and abolished at the end of 1965.[85] Since Venezuela was the traditional supplier of residual fuel to Atlantic Coast states, these allowances favored the South American nation without actually saying it. To ignore Venezuelan complaints about discrimination would, in the Department of State's view, "create a situation of maximum risk of nationalistic reaction" while "a concession of this kind will weaken the position of those Venezuelans who assert that our oil policy disregards Venezuela's needs."[86] In any case, the concessions dovetailed with U.S. refiners' desires to produce products other than cheaply priced residual fuel.

U.S. officials were opposed, however, to Pérez Alfonzo's goals. To grant Venezuela preferential treatment would only risk angering other oil exporters, like Iran and Saudi Arabia. Moreover, while they had not protested, officials were unhappy about Venezuela's new taxes on the oil companies. An exemption from the quotas would, the State Department believed, "appear to be rewarding nationalism and thereby encourage nationalism." As for the proposal that the government purchase Venezuelan oil and help control production and set prices, this was "contrary to United States interests." The National Security Council warned "that the granting of an exemption or preference would be interpreted as a tacit United States recognition that the theories outlined above could have been advantageous to Venezuela in practice; and it would therefore increase the risk that these theories might be tried out in Venezuela or elsewhere." The State Department agreed that Pérez Alfonzo's ideas touched more than Venezuelan-U.S. relations. If the Venezuelan succeeded, the results "would not be limited to the possible unsettling effect on American investment and interests in Venezuela and Latin America," for "the position of our oil companies throughout the world would be weakened, as would the assured access of consuming nations to adequate supplies of oil at reasonable prices."[87]

To be sure, when the Eisenhower, Kennedy, and Johnson administrations entertained repeated Venezuelan requests for relief from the quotas, they were not free to pursue the national interest apart from political considerations. A variety of special-interest groups pushed and pulled at the oil import program, causing it to be amended seven times between 1959 and mid-1965. As Bureau of the Budget Director Charles Schultze related in exasperation, "the result is an exceedingly complex legal foundation for a program that involves broad and competing considerations of domestic and foreign policy, including national security, foreign trade, antitrust laws, consumer interests, and protection and expansion of the domestic petroleum industry."[88] Along with oil producers and their political allies, coal-mine operators and miners wanted restrictions on Venezuelan residual fuel which, they alleged, displaced coal. The tough-talking president of the United Mine Workers, Anthony "Tony" Boyle, charged "that the residual fuel oil levels were set to benefit Venezuela" and that "America came first." Yet, Venezuelan petroleum had its U.S. friends. Free traders, skeptics of national security arguments, and public officials from oil-importing areas like New England decried the oil import program and discrimination against Venezuela.[89]

The interests that spoke loudest in the debate over U.S. oil policies toward Venezuela, however, were the multinational oil companies, particularly Standard Oil of New Jersey and Gulf Oil. While ambivalent about import quotas, since they ideally preferred an open world, they firmly opposed any preferential treatment for their host. The companies never forgave Venezuela for the tax increase at the end of 1958, and they demanded that, before considering a preference, Venezuela be forced to freeze taxes, disband the national oil company, and discard ideas on price and production controls. As Gulf wrote to the National Security Council, "Venezuela must be made fully aware that it must assume full responsibility for restrictive measures taken to meet immediate and limited objectives or for narrow nationalistic aspirations." Industry leaders also warned that governments that pursued "statist oil policies weaken the entrepreneurial group that opposes communism."[90] Increasingly through the 1960s, some officials in the Johnson administration, alarmed over

Venezuelan bitterness, recommended some form of preference. But the oil companies, according to Ambassador to Venezuela Bernbaum, blocked any significant change; they were determined to teach Pérez Alfonzo the "facts of life concerning the international oil business," and they feared the effect on their investments in such other areas as the Middle East. As Bernbaum recalled, "the oil industry realized they were playing a dangerous game," but "they still insisted on playing it" on the grounds that "they are the ones who have the interest, they are the ones who pay for it—they face the risks."[91]

While wielding impressive power, Standard Oil and Gulf Oil did not dictate U.S. foreign policy. Even if Washington had granted Venezuela parity with Canada, it still would have rejected both the Prebisch thesis on relative prices and Pérez Alfonzo's notion of the intrinsic value of petroleum.[92] When, for example, the Leoni government tried to take advantage of an expansion of the residual fuel quota by insisting that the companies raise the price of residual fuel, the Johnson administration reacted sharply, expressing "serious concern with any actions that would result in an increase in the price of residual fuel to the U.S. consumer."[93] During the 1960s, the United States remained firmly committed to its traditional belief in the efficacy of the free exchange of goods and services, albeit it violated its own principles with quotas on oil imports. It also, of course, believed that private interests should control world trade, although, by defending the quotas on national security grounds, it seemed to suggest that petroleum could not be equated with peaches and pineapples. But world prosperity depended upon economic growth and the expansion of U.S. investments abroad, not prorationing schemes, intergovernmental deals, or the indexing of the prices of raw materials to the prices of manufactures. Pérez Alfonzo's plans challenged the principles embodied in the Open Door, the foreign economic policy of the United States. As Lyndon Johnson explained in August 1967 in a letter to President Leoni, the United States wanted to help the Venezuelan economy, but "all policies and alternatives must be measured by their political and economic feasibility."[94]

The repeated U.S. rebuffs of Venezuela's position on oil gradually estranged relations between the two nations. In part, Venezuela resented the preferences granted Canada and Mexico. By excluding Venezuelan oil from its national security program, the United States shattered the national myth of a small, poor nation that its contributions had been essential to the Western world during past conflicts. As Ambassador Bernbaum remarked, Venezuelans felt discriminated against in a "psychological" sense.[95] After the Arab-Israeli war of 1967, when Arab nations embargoed oil shipments to the United

States for three months, Venezuelan leaders were convinced that Washington, appreciative of Venezuela's 350,000-barrel-a-day increase in production, would finally judge their oil as "secure" and no longer "lump it with oil from sources outside the Hemisphere." Instead, the White House and oilmen, pointing to a one-million-barrel-a-day increase in domestic production, cited the U.S. ability to shrug the embargo as proof of the wisdom of import restrictions. Then Venezuela was criticized by its Arab friends for undermining the embargo.[96]

The crux of Venezuelan protest, however, was economic, and it involved the Alliance for Progress. "No matter how big the funds assigned to the program are," President Betancourt told newsmen in December 1963, "they will never compensate for the damage done by what economists call 'trade terms' betwen Latin America and the industrialized countries." The alliance called for long-range economic planning, but, without price stability for key exports, "it will be next to impossible for Latin America to overcome its present difficulties and set its course toward sound development."[97] Particularly during the Leoni administration, with the United States also curtailing economic aid, Venezuelan criticism became strident. In a letter to President Johnson and in an article in the influential journal *Foreign Affairs*, Leoni accused the United States of inflicting "serious damage to the Venezuelan economy." He also protested by declining an invitation to visit Washington.[98] Venezuela had lost faith, so evident in the early 1960s, that the United States was truly committed to improving the lives of Latin Americans.

At least for the decade, Venezuela's new oil policies were failures. Supervision of the foreign oil companies remained limited, OPEC was in disarray, and the price of oil declined. The nation's major achievement, enhanced by new tax measures in 1966, was to collect approximately two-thirds of the oil companies' profits. But, due to the slide in prices, the government actually realized less money per barrel than it did during the Pérez Jiménez years; in 1957, for example, the government received $1.03 per barrel, while in 1967 it received 98.3¢ a barrel.[99] As such, to increase income the government had to depend on the traditional method of expanding production of a nonrenewable resource. Production rose slowly and unevenly from slightly over 1 billion barrels in 1957 to 1.3 billion barrels in 1968. This gave the Leoni administration approximately $300 million more in oil revenue to spend during its last year in office than Pérez Jiménez had in the year before his overthrow.[100]

In urging a bold program for Latin America, presidential advisor Schlesinger told John Kennedy that Latin America was "set for miracles." Even allowing for bureaucratic inefficiency and wasteful spending on the military, the Acción Democrática governments lacked the money to perform marvelous feats. What they did was improve the lots of common people by cutting unemployment in half and allocating one-fourth of the budget to health and education. They also resettled approximately 160,000 families on their own farms. While these programs surpassed efforts made in most countries, they did not produce a prosperous, egalitarian society. The effects of the population boom, over 3 percent growth a year, kept the nation from hitting the alliance's target of a real economic growth of 2.5 percent a year; per capita gross domestic product barely rose during the decade. Over one-half of the rural population continued to live as subsistence farmers, and the *ranchos* proliferated as the population of Caracas grew to nearly two million by 1969. Three out of every four children dropped out of school by the sixth grade, because frequently they worked to help their families survive. In sum, at least one-half of the population was still poorly fed, clothed, and housed.[101] In their moderate, evolutionary ways, the Betancourt and Leoni governments worked diligently for the poor, but only stable prices and reliable markets for Venezuelan oil would give the government the resources necessary for its war on poverty. As Leoni told the foreign-policy establishment of the United States, "frankly, loans and credits are not as desirable as fair prices."[102]

The Acción Democrática governments were also unable to alter significantly the structure of the Venezuelan economy. Oil still accounted for 90 percent of the value of its exports and approximately two-thirds of the treasury's revenue.[103] The country's trade continued along traditional lines. It sold most of its coffee and iron ore to the United States and, despite the quotas, about 40 percent of its petroleum production to its northern neighbor. In return, Venezuela bought more than one-half of its imports from U.S. merchants, as the reciprocal trade agreement of 1939 remained in effect.[104]

This persistence in the patterns of Venezuela's economy suggested that Venezuela continued to be a lucrative market for U.S. traders and investors. With annual sales as high as $780 million, U.S. exporters counted Venezuela the third best market in Latin America and eleventh in the world. Total direct investment stood in 1970 at $2.7 billion, still the highest investment in Latin America. After relatively poor performances in the early 1960s, the oil companies belied their claims that Venezuelan oil was overtaxed by making a return on the value of their net fixed assets that reached as high as 39.5 percent in 1968. The steel companies, according to government data, did well also, as their returns on net fixed assets went over 30 percent twice during the 1960s.[105]

From the perspective of Washington, Venezuela represented one of the few successes of the Alliance for Progress, a program widely considered by the end of the 1960s to have failed. The South American nation, with U.S. assistance, took important steps toward the alliance's objectives of political democracy, economic development, and social reform. In addition, the Acción Democrática governments thrashed the political left. Yet, the end of the decade saw Venezuela still fundamentally a poor nation and its leaders exasperated with the United States. While both nations favored orderly, progressive change, Venezuela no longer believed that liberal economic principles coupled with foreign aid from the United States would solve its basic problems. Pérez Alfonzo's critique of the international economic system implied a structural change in the relationship between industrial nations and raw-material producers. While eager to help Venezuelans, the United States never intended to redistribute the world's wealth with the Alliance for Progress.

8. The Energy Crisis, 1969–1976

Between 1969 and 1976, relations between the United States and Venezuela underwent a historic change. Venezuela no longer had to rely on U.S. investors or the goodwill of the United States for its prosperity. In 1973, thirteen years after its formation, the Organization of Petroleum Exporting Countries (OPEC) became a powerful cartel which arbitrarily set the price of a barrel of oil. The cartel's audacious policy succeeded because the bargaining positions of the United States and the oil producers had been fundamentally altered. An "energy crisis" struck; OPEC's actions coincided with the U.S. discovery that domestic oilfields could no longer produce enough fuel to drive the economy. Foreign supplies were essential. Emboldened by the U.S. energy crisis and conscious of their new power, Venezuelan leaders denounced the old economic relationship with the United States, nationalized U.S. oil and steel holdings, and fashioned a foreign economic policy that differed from Washington's. For the first time in the twentieth century, the United States was unable to control or channel Venezuelan nationalism. Seemingly unwilling to curb its voracious appetite for petroleum, the United States had lost some of its diplomatic clout in the Western Hemisphere by ironically becoming partially dependent upon a Caribbean nation.

New presidential administrations assumed power in early 1969 in both Caracas and Washington. The Venezuelan president was Dr. Rafael Caldera, the perennial head of the Christian Democratic party, COPEI. The inauguration of a Christian Democrat was another benchmark in Venezuela's political history. The presidential sash had not only been peacefully transferred again but also passed for the first time to someone other than a leader of Acción Democrática. Though they regarded themselves as the authentic representatives of the people, the *adecos* relinquished power. Indeed, Raúl Leoni, the former president, lectured his party on the virtues of loyal opposition in sustaining a democracy when some *adeco* congressmen refused to cooperate with the new administration. Acción Democrática's grudging acceptance of the minority COPEI government strengthened civilian rule and ensured that a one-party system, like Mexico's, would not dominate Venezuelan life.[1]

Venezuela's third popularly elected president since 1958 was born in 1916 in San Felipe in the north-central state of Yaracuy. A man of letters, Caldera had earned a doctor of philosophy degree in political science, taught at the Universidad Central, and written a scholarly study of the distinguished Latin American educator Andrés Bello. His scholarly interests also included the rights and responsibilities of labor. He served in the Ministry of Labor between 1936 and 1938 and published a comprehensive review of Venezuela's labor laws. Caldera's political career began in 1936 when he was elected to the Chamber of Deputies. During the political ferment of the *trienio*, Caldera, a devout Catholic, founded COPEI. Initially jealous of Acción Democrática's hold on the Venezuelan polity, neither he nor his party opposed the military *golpe* against the Gallegos government in 1948. But, as the Pérez Jiménez tyranny stretched through the 1950s, Caldera began to denounce the dictatorship and went into exile. Like the *adecos*, Caldera concluded that civilian politicians needed to cooperate with each other; he signed the Pact of Punto Fijo and supported the Betancourt administration during the turbulent period between 1959 and 1963. His party's position enhanced by elections in late 1963, Caldera, citing official ineptness and corruption, peacefully but vigorously opposed the Leoni government. After repeated bids for the presidency, the Christian Democrat's persistency was rewarded when he garnered 29 percent of the vote in a multicandidate election in December 1968.[2]

As a Christian Democrat, President Caldera pledged to care for the moral and spiritual needs as well as the material aspirations of Venezuelans. And as an anti-Marxist party with backing from business and propertied groups, it seemed that the Caldera government would slow the pace of change in the nation by granting a larger role to private enterprise in Venezuela's development. The foreign oil companies, for example, were cheered by Caldera's victory, pleased to be rid of both the party that raised their taxes and their inveterate antagonist, Juan Pablo Pérez Alfonzo.[3] In practice, Caldera's administration differed little from its predecessors because, in part, it lacked a ruling majority in Congress. It too pursued moderate, evolutionary reform, working through entrenched power groups and within the framework of constitutional democracy. Economic development and growth continued to rest upon governmental expenditures, which ultimately depended upon revenues generated by the

U.S. and British-Dutch oil companies. What distinguished Caldera's rule from those of Betancourt and Leoni was a growing confidence in Venezuela's decade-old political institutions. The new president declared an amnesty for the few remaining leftist insurgents, provided they agreed to respect democratic processes. Caldera also renounced the Betancourt Doctrine; he judged that Venezuela's recognition of extraconstitutional changes of power in other countries would not encourage the Venezuelan military to strike another *golpe*. Finally, he demonstrated that Venezuela no longer felt threatened by others when he declared his intention to establish relations with socialist nations like Cuba and the Soviet Union.[4]

President Caldera's conciliatory gestures toward former foes did not signify that he believed peace and stability were assured, but rather that he considered the survival of Venezuelan democracy more an economic question than a political one. Throughout his administration, Caldera repeatedly called for "international social justice," or a world where "every country may have access to the means to achieve its own development." The poor nations did not want financial assistance, or welfare, from the industrial world; they wanted expanded markets and stable prices for their exports and access to technological and scientific advances so that they might finance their own economic growth. For Venezuela, this meant, of course, fair prices and open markets for petroleum. Caldera grounded his "fervent advocacy of international social justice" on basic Christian principles of "justice . . . and for the sake of the common good," as espoused, for example, by Pope John XXIII.[5] His thesis was consistent, however, with the ideas of Pérez Alfonzo, Raúl Prebisch, and the position adopted by Third World nations in the 1960s at such forums as the United Nations Conference on Trade and Development (UNCTAD). The battlecry of the Third World would be a "New International Economic Order," or a relationship in which the prices of minerals and commodities exported by nonindustrial nations kept pace with the prices of industrial wares.[6]

The call for international social justice was met indifferently by the presidential administration of Richard M. Nixon. At least until the end of 1973, Nixon and chief foreign-policy advisor, Henry A. Kissinger, assigned a low priority to relations with Latin America. Their major international concerns were to extricate U.S. troops from Indochina while attaining "peace with honor," preventing a "superpower" confrontation in the Middle East, and constructing a "generation of peace" through understandings or détente with the Soviet Union and the People's Republic of China. As such, the Nixon government lumped Latin America, except perhaps for Brazil, with other poor, weak Third World nations who seemingly weighed little in the balance of power. The United States had not, however, renounced its suzerainty in the Western Hemisphere. The election in 1970 of Salvador Allende, a Marxist, as president of Chile evoked a sustained series of clandestine operations by the United States to unseat that government. But publicly, the Nixon administration's policy toward Latin America was to project a "low profile," rejecting the advice of its own fact-finding mission to Latin America headed by Nelson Rockefeller to establish new regional alliances and programs. President Nixon told correspondents from Latin America that the United States would no longer pursue "the illusion that we alone can remake continents."[7]

With its emphasis on strategic security and the balance of power, the Nixon government gave scant attention to Latin America's new economic concerns. In May 1969, at an inter–Latin American conclave, delegates produced the "Consensus of Viña del Mar." The conference, which was intended to revivify the moribund Alliance for Progress, essentially adopted the same program contained in Rafael Caldera's "international social justice." The delegates called "for a fairer international division of labor that will favor the rapid economic and social development of the developing countries, instead of impeding it as has been the case hitherto." Trade, not aid, was what Latin America wanted. When loans and credits were necessary, they should be negotiated through multilateral lending agencies, like the Inter-American Development Bank, because bilateral loans from Washington, the delegates bluntly noted, had too many conditions "tied" to them.[8] At least rhetorically, the Consensus of Viña del Mar received a favorable response from the Nixon administration. President Nixon pledged to continue assistance, "untie" that aid, and reduce tariff and nontariff barriers against Latin American exports to the United States. Little effort was exerted, however, to convince Congress to implement these recommendations. Indeed, the United States sharply reduced economic assistance; aid to Latin America in 1971, for example, was less than $500 million, 50 percent less than in most years of the 1960s. And conditions were still attached to development assistance. Aid recipients were not permitted to spend their grants in Western Europe or Japan even though these areas might sell the desired capital goods and sophisticated equipment at a favorable price. In addition, Washington imposed new tariff restrictions, such as the 10 percent surcharge placed on all imports in late 1971.[9]

The Nixon administration's attitude toward Latin America was reflected in its diplomacy toward Venezuela. Relations between the two new governments began sourly. Caracas, fearful of a repetition of the Nixon incident of 1958, canceled Nelson Rockefeller's visit in June 1969, just hours before he was to arrive in Venezuela. Rockefeller, who had witnessed anti-U.S. demonstrations in eight Latin American capitals during his tour, claimed that the cancellation was "the result of a coordinated effort by highly organized and militant forces receiving direction from inside and outside the hemisphere." Whatever the merits of Rockefeller's analysis, the Caldera government was not eager to entertain him since it wanted a comprehensive, systematic review of Venezuelan-U.S. relations, instead of an exchange of views.[10] The Venezuelans were similarly reluctant to welcome another U.S. visitor, John G. Hurd, President Nixon's choice to be ambassador to Venezuela. Hurd, who had headed Nixon's election campaign in Texas, was a prominent member of the Independent Petroleum Producers Association of America, the very organization that lobbied for import quotas on Venezuelan oil. Nixon withdrew Hurd's nomination from the Senate in August 1969 after some senators objected. Frank Church of Idaho, the chairman of the Subcommittee on the Western Hemisphere of the Foreign Relations Committee, remarked, for example, that "to send an oil protectionist to Caracas is like sending a Zionist to Cairo."[11] Perhaps piqued by the Venezuelan attitude, the administration waited a year before sending Robert McClintock, a career foreign officer and associate of Rockefeller, to be its representative in Caracas.

The issues that awaited Ambassador McClintock were the same as those of the 1960s: the U.S. import quotas, the price of petroleum, and the behavior of the foreign oil companies. The import quota question arose immediately, for President Nixon appointed a cabinet-level task force to review the oil import program, which had been repeatedly amended since 1959. The task force consisted of the secretaries of state, defense, treasury, commerce, and interior with Secretary of Labor George Shultz serving as chairman. Its executive director was Philip Areeda, a professor of law at Harvard University and a member of the National Security Council during the Eisenhower administration. The task force solicited the views of interested parties. The debate had not shifted since 1959; domestic oil producers, warning that "the military, economic, and political security of the United States requires that we not become dependent on overseas energy supplies," argued for retention of quotas. Consumer groups answered that U.S. citizens had unnecessarily paid an extra $5 billion to subsidize the oil industry.[12]

Crucial to the resolution of the debate was an accurate analysis of the oil reserves of the United States. In 1968, domestic production of crude oil and natural gas liquids, a daily output of over ten million barrels a day, met approximately 80 percent of the demand. Trade groups, like the American Petroleum Institute and the Independent Petroleum Producers, predicted that, with continued protection from foreign oil, they would be able to expand domestic production by perhaps as much as 33 percent to satisfy growing demands for petroleum products. The multinational oil companies questioned, however, the ability of the domestic industry to expand production. Standard Oil of New Jersey, reversing its stand of the 1960s, even suggested to the task force that Venezuelan oil be given special consideration, although the company did not specifically recommend a preference and certainly opposed government-to-government negotiations.[13]

The task force also took testimony from Venezuela. The new minister of mines, Hugo Pérez de la Salvia, traveled to Washington to press his nation's case, and his brief was supplemented by a series of speeches by the ambassador to the United States, Julio Sosa-Rodríguez. Venezuela again requested preferential treatment for its oil in the U.S. market or at least parity with Mexico and Canada, for our "economies are intimately linked as a matter of tradition, geographical location, and mutual interests." They reminded their hosts that Venezuela was a reliable supplier, having sold oil to the United States for over forty years. And they pointed out the discriminatory effects of the oil import program: the bulk of U.S. purchases were for residual fuel oil, which was priced 16 percent below crude oil. Failure to achieve a secure market, Sosa-Rodríguez feared, "would endanger the fundamental basis of our development and would lead to a loss of faith in the democratic system as the irreplaceable formula for the progress of people." The task force's report would also determine, in President Caldera's words, whether the United States considered Venezuela a "second-class friend."[14]

The cabinet's report, which was released in February 1970, failed to settle the oil import controversy. Attempting to promote the national security while addressing the concerns of domestic producers, consumers, Canada, and Venezuela, the secretaries of state, defense, and treasury recommended that the import quota system be scrapped and replaced with a multitiered tariff. Overland shipments of crude from Canada and Mexico and seaborne imports of residual fuel oil, which was not produced in sufficient quantities by U.S. refineries, would be allowed to enter duty free. A tariff of seventy cents a barrel would be imposed on imports of crude oil from Venezuela, while imports from the Eastern Hemisphere would carry a duty of ninety cents a barrel. The secretaries judged that this tariff system would protect domestic producers from unfair price competition but would still spur them to produce economically, since they could no longer hide behind arbitrary limitations on the volume of oil imports. The national security would also be enhanced by the new program because it gave tariff preferences to Canadian and Latin American oil. The State Deaprtment did not want the United States to depend heavily on oil imports from the politically turbulent areas of Africa and the Middle East, which in 1968 supplied 13 percent of U.S. imports or about 3 percent of total consumption. Instead, the United States should rely on Venezuela, which, the secretaries noted, had historically been a dependable supplier and was politically stable. Moreover, the Department of Defense deemed Venezuelan oil no less secure than seaborne domestic oil because "the Caribbean and coastal waters affected during World War II are now significantly more secure than the high seas."[15]

While the cabinet's report partially satisfied Venezuela by continuing the unrestricted entry of residual fuel into the United States and by differentiating between Venezuelan and Middle Eastern crude, it denied the intent behind Pérez Alfonzo's original request for a preference and President Caldera's call for international social justice. The price of petroleum had declined by 25 percent during the 1960s, while the prices of industrial wares, Venezuelans alleged, had risen by 25 percent. The terms of this trade, Caldera charged, "constituted one of the most accusing documents that might be written in the history of mankind."[16] While admitting that Venezuelan oil was secure, the cabinet refused to grant Venezuela parity with Canada. It ventured that the South American nation, which supplied about one-half of the U.S. imports, could use a preference to expand its imports by underselling U.S. oil, which sold for more than a dollar a barrel above the Venezuelan price. But, since they had persistently objected to low prices, the Venezuelans probably would sharply raise taxes on the foreign oil companies so that the price of their petroleum would equal that of U.S.-produced oil. Making Venezuela a full partner in this national security program was not worth, in the task force's opinion, leaving control over prices and production "in Venezuelan hands." The report recommended that the effect of the proposed tariff on Venezuela be reviewed in 1975 and predicted that Venezuelan oil might enter duty free in the 1980s, when demand for petroleum grew in the United States.[17]

The task force's work, which was the last extensive review of oil policy prior to the Arab-Israeli War of 1973 and the ensuing oil embargo, proved fruitless. President Nixon sent the report without comment to Congress; in effect, he postponed a decision on it for a year. To erect the multitiered tariff would be politically sensitive, for it would anger oil-producing states like Texas and Oklahoma, whose support Nixon might need in his re-election campaign. Indeed, the secretaries of commerce and interior, whose constituencies were more national than international, had dissented from the majority recommendation to abolish the quotas. In the House of Representatives, the report received a frigid reception from both the Ways and Means Committee and the Mining Subcommittee of the Interior Committee, two bodies in which oil interests wielded significant influence. Bowing to the oil lobby's pressure, Nixon announced in August 1970 that he would retain the quotas, thereby overruling his own cabinet.[18] The effect of his decision was to deny Venezuela any special consideration, as it had been repeatedly denied since 1959. This time, however, domestic producers, not the multinational oil companies, worked feverishly to block change, since the issue was the quota, not, as in the 1960s, an exemption from the quota for Venezuela.

Faced with a distraught Venezuela, the Nixon government responded as had the three previous administrations. In June 1970, in the midst of the congressional debate over the task force's report, President Caldera journeyed to Washington with the mission of securing "price and volume stability, together with an adequate growth rate, for Venezuelan exports to the United States." In speeches, including one to a joint session of Congress, Caldera presented his nation's case within the context of his vision of international social justice.[19] What he left with was a 35,000 barrel a day increase in the U.S. quota on home heating oil and President Nixon's observation on the need for a hemispheric policy on oil. Until early 1973, the Caldera government beleived that the United States would eventually recognize "the justice of our position and of the advantages which hemispheric treatment would have for both countries." Instead, it received more piecemeal concessions on various petroleum products.[20]

What small chance that President Nixon would draw upon the national security provisions of the Trade Act and exempt Venezuela from the oil quota was doomed by his administration's reaction to new Venezuelan laws regulating the U.S.-owned oil industry. Though its oil officials were perhaps less articulate and possessed less of a sense of history than Pérez Alfonzo did, the Caldera administration shared his goal of gaining control over the nation's economic destiny. If it wavered, congressional friends of the retired oil minister reminded the Christian Democrats of their duty. Moreover, like the Acción Democrática governments, the administration needed expanding tax revenues to create each year 120,000 new jobs for the burgeoning population, which was growing at an annual rate of over 3 percent. Between 1969 and 1973, the government raised taxes, set production goals for the companies, and ordered them to post bonds on their property, ensuring that, when concessions obtained under the oil law of 1943 reverted back to the government in 1983, the oilfields and machinery would be in satisfactory condition.[21]

As they had since the fall of Pérez Jiménez, the oil companies tenaciously resisted nationalistic measures. Leo Lowry, president of Creole, decried the "official intervention in the oil business" and warned that the government was creating "a serious impediment to investment." Another industry spokesman flatly predicted that "Venezuela could tax itself right out of the oil business." Beyond objecting vigorously, the companies opposed the government by continuing to allow oil reserves to dwindle and perhaps by deliberately cutting production. Venezuela's output hit a historic high of 1.35 billion barrels in 1970 but declined by 4 percent in 1971 and by another 9 percent in 1972. The companies claimed that mild winters in the northeastern areas of the United States and the uncompetitive price of Venezuelan oil explained the drop in production. Venezuelan officials charged, however, that the companies were boycotting their oil in reprisal for the new taxes.[22] Whatever the cause, the decline in production curtailed the government's spending plans and contributed to a budgetary deficit in fiscal year 1972.

Oilmen complained that Washington was indifferent to this "de facto nationalization" and "plain expropriation" of their holdings.[23] While the Nixon administration, faithful to its low-profile approach, did not publicly denounce Venezuela, it relentlessly pressured Caracas to withdraw the new taxes. In February 1971, in a significant speech, Ambassador McClintock strongly implied to the Venezuelan Association of Executives that the Caldera government was jeopardizing the nation's economic health. The United States, McClintock told the businessmen, had welcomed foreign investment in the nineteenth century, when it was a developing nation. That investment aided the United States just as "on balance foreign investors have certainly created more wealth in Venezuela than they have taken out." Complaints about terms of trade were exaggerated, for statistics on trade and balance of payments disguised the benefits of foreign investment, ignoring money spent and taxes paid in the host nation. McClintock reminded them that the oil and steel companies had paid $10.3 billion in taxes in the 1960s. The ambassador noted that Venezuela wanted a preference in the U.S. market, but he wondered whether, in view of the new impositions on the oil companies, "investment capital can be found in sufficient quantity to keep up future production, which alone can provide security of source." This reaffirmation of the traditional foreign economic policy of the United States ended with the prediction that cooperation with U.S. capital would contribute to Venezuela's "attainment of peace and prosperity within the framework of democracy."[24]

When persuasion failed, the Nixon administration issued stern warnings to the Venezuelans. After the Venezuelan Congress passed in early 1972 the Reversion Law, which required the companies to bond their properties, the government reportedly "let it be known unofficially that the U.S. would be looking towards Canada rather than Venezuela in the future to meet its demand for oil."[25] Officially, it sent John Connally, one member of the administration who advocated getting "tough with Latin America because we don't have any friends left there anyway," to Caracas to object to "nationalistic tendencies on foreign investment." The government backed these warnings by deciding in September 1972 not to give Venezuela any special share of the latest 600,000 barrel a day increase in the quota.[26]

One final episode before the outbreak of the Arab-Israeli War of 1973 demonstrated how sharply different were the views of the United States and Venezuela on the issues of investment and trade. The greatest potential source of petroleum in the Western Hemisphere is the "heavy oil" zone in the southeastern part of Venezuela along the northern bank of the Orinoco River. This belt contains perhaps a staggering 700 billion barrels of molasses-like petroleum, which is also laden with heavy metals and sulfur. Geologists and petroleum engineers believe that, with technological advances and massive infusions of capital, a significant amount of the Orinoco Tar Belt can be commercially recovered.[27] A few perspicacious Department of State officials, who foresaw a "serious energy crisis by 1980," began in 1972 to urge that the United States approach Venezuela about developing the tar belt. President Caldera welcomed the initiative, hoping that his country would receive by treaty the investment capital and guaranteed markets and prices that would secure his country's future. Department officials discussed with Venezuela the free entry of heavy oils into the United States "in return for investment guarantees to the companies developing these oils." But Venezuela wanted a government-to-government arrangement, with the United States paying for the new technology; it was not interested in dickering with private oil companies. The U.S. position, enunciated in July 1973, was, however, that bilateral "deals" would set off a bidding war, pitting one consuming nation against another.[28] Perhaps in a less turbulent period the two countries might have eventually negotiated a service-contract arrangement, with the State Department persuading the oil companies to work for Venezuela, the Latin American government paying them a fee for their services, and the United States guaranteeing a stable market and high prices for heavy oil products. Certainly, some department officers, particularly those stationed in Caracas, such as Chargé d'Affaires Francis Herron, argued for a fundamental revision of oil policies. But, as one scholar put it, on the very eve of the "energy crisis," the United States had not recognized "the futility of trying to salvage private multinational control of mineral resources in developing countries."[29] Indeed, as a Senate committee concluded in 1975, the United States had traditionally assumed that the interests of the nation and the multinational oil companies were identical.[30]

Why the United States declined to refashion its diplomacy toward Venezuela can be explained not only by its faith in free trade and investment principles but also by its ignorance of the rapidly changing world of oil. Two simultaneous developments, the failure of domestic oil companies to meet anticipated production goals and the unanticipated ability of OPEC to act cohesively, left the United States vulnerable to an energy crunch. The United States began the 1970s confident both of its own and the noncommunist world's oil supplies. While it had been a net importer of oil since 1948, the United States had enjoyed a steady increase in its domestic output, particularly during the 1960s. Crude oil production in 1970 was 3.52 billion barrels, 27 percent higher than in 1960. Production increases in the Middle East and Africa, especially Libya, were, of course, stupendous. Between 1965 and 1970, production in these areas doubled to an output of 7.31 billion barrels. Then new areas, such as Alaska, the continental shelf, and the North Sea, would soon be producing oil, and other areas, perhaps Australia and the tar belts in Venezuela and Canada, presented good possibilities. In view of past developments and future prospects, the task force on oil imports predicted in 1970 that the United States would need to import only about 25 percent of its oil needs and that "the landed price of foreign crude by 1980 may well decline and will in any event not experience a substantial increase."[31] Prominent industry spokesmen and independent oil analysts supported the task force's conclusions. For example, Michael L. Haider, chairman of the board of Exxon, assured his stockholders that the world's oil reserves would not be depleted even "in the lifetimes of our grandchildren."[32]

These optimistic estimates were wrong, both on the supply and the demand sides. Domestic production peaked in 1970 and then steadily declined. The task force predicted, for example, that in 1975 domestic output of crude oil would be 3.94 billion barrels; it was 3.05 billion barrels. Production failed to meet anticipated goals because the northern slope of Alaska and the continental shelf proved not to be bonanzas and because estimates of domestic production were based on figures supplied by the oil industry. These may have been self-serving statistics, for the industry wanted to justify the retention of quotas on foreign imports.[33] Once the energy crisis became apparent, this lack of independent information on oil reserves would, of course, feed the widespread public suspicion that the industry deliberately withheld supplies to raise prices. Estimates on production in Canada and Venezuela, where the United States planned to purchase its imports, were also wildly wrong. The task force spoke of Venezuela virtually doubling output in the 1970s, but there, too, production peaked in 1970. Within a few years, the contribution of the Western Hemisphere to world oil production would dramatically decline. In 1970, the United States and Latin America produced one-third of the world's oil; in 1973, they accounted for only about one-fourth of production.[34]

While supplies were declining, demand rose more rapidly than expected. In 1968, for example, the United States used about 13.2 million barrels of petroleum a day, but five years later, in the days before the oil embargo, about 17 million barrels a day were consumed. Demand soared because, for numerous reasons, U.S. citizens chose to use oil instead of other sources of energy, such as coal, natural gas, and nuclear power. World demand similarly rose, for northern and western European nations also consumed more petroleum as they achieved standards of living that matched and even surpassed that of the United States.[35]

The key development in the swiftly evolving oil business was that after 1970 U.S. oilfields were probably working at maximum capacity. Owning now less than 10 percent of the world's known reserves, the United States no longer had the spare capacity to raise production to defeat an embargo, as it did in 1967, or to undercut any fixing of the price of oil.[36] Even during the mid-1960s, when OPEC was considered impotent, international oil analysts warned that "the importance of ample U.S. oil reserves and excess capacity is immense—not just for direct national defense in the narrow sense but in terms of our worldwide power position." If the United States became "decisively dependent" on foreign oil, it would become subject to "unacceptable" pressures from oil-producing countries. Thus, "it is the U.S. surplus crude capacity which provides the essential protection for the oil economy of the Free World."[37]

How OPEC discovered and then seized upon the U.S. weakness is an intricate and complex story. But certain developments and events stand out. The embarrassing defeat suffered at the hands of Israel in 1967 imparted a sense of unity and purpose, which was lacking in the initial OPEC meetings, among Arab oil producers. In particular, that military debacle and the virtually unquestioning U.S. support of Israel caused the conservative, staunchly anticommunist monarch of Saudi Arabia, King Faisal, to waver in his traditional friendship for the United States. The head of the other major oil producer in the Middle East, Shah Mohammed Reza Pahlevi of Iran, was also growing restless about his status as a client of the West. The Iranian monarch wanted to play an expanded diplomatic and military role in the Middle East and southwestern Asia. These kingly aspirations seemed feasible, for the producers of the Middle East and North Africa were developing through experience and the technical assistance of Venezuela an understanding and knowledge of the international oil trade. Oil ministers, like Saudi Arabia's Sheik Ahmed Zaki Yamani, no longer felt compelled to accept routinely industry explanations and reports. OPEC members also gradually adopted Third World themes that industrial nations were dictating the terms of trade and that foreign control of mineral resources was a vestige of colonialism.[38]

The catalyst for change was the fiery pan-Arabist and virulently anti-Zionist leader of Libya, Colonel Muammar el-Qaddafi. Qaddafi, who deposed Libya's corrupt, pro-Western monarch in 1969, demanded and secured in September 1970 a revision of Libya's oil contracts and thirty cents more a barrel from Occidental Petroleum Company. Beyond fortuitous timing, Qaddafi succeeded because he seemed fanatical enough to make good his threat to shut down Libya's 3.3 million barrel a day production. As he proclaimed, "the people who have lived 5,000 years without petroleum can live without it for many more decades in order to achieve their legitimate rights." The Libyan's victory initiated three years of "Libyan leap-frog," as individual OPEC members repeatedly pressed the multinational oil companies for new price and profit concessions and a share in their operations. The price of a barrel of crude, which was below $2.00 a barrel in the 1960s, rose quickly. Venezuela, for example, ordered the companies to raise prices four times, up to $4.44 a barrel, between January and October 1973. The companies, because of tight supplies, were unable to resist these demands by the traditional method of playing one country off against another. As Sheik Yamani remarked during negotiations with Exxon officials, "you know the supply situation better than I. You know that you cannot take a shutdown." Even without the October War and the Arab oil embargo, the price of oil would have continued to soar. OPEC planned to meet in October 1973 and discuss a price of between $4.00 and $6.00 a barrel. The chaos generated in the West by the embargo, which began on October 18 and lasted until March 18, 1974, simply enhanced OPEC's sense of power. Led by Iran, the oil cartel set on December 22, 1973, a price of $11.65 a barrel.[39]

The United States failed to devise an effective strategy to cope with the developing strength of OPEC. Some State Department officials, led by energy expert James Aikins, began in 1969–1970 to warn that U.S. oil policies were based on inaccurate and misleading estimates on the supply and demand of oil. But their gloomy forecasts about prices and the growing dependence of the West on the Middle East raised little alarm. Believing "that the United States was becoming vaguely hysterical as its import needs grew," the Europeans, who traditionally relied on imported oil, ignored U.S. suggestions to develop a common strategy. Moreover, the oil companies apparently believed that they could curb OPEC's appetite without the intercession of the State Department.[40] In any case, to listen to Aikins would lead to a dramatic shift in U.S. foreign policy, for he concluded that Arab nations could control oil supplies and prices. But not until after October 1973 was either the U.S. public or its elected officials willing to weigh aspirations of Arab peoples against traditional support for Israel. And if all understood, as Under Secretary of State John Irwin told Congress in 1972, that "petroleum is a finite commodity that is essential to an extent unequaled by any other commodity to a country's well-being and that it is the most political of all commodities," it is still doubtful whether the United States would have overruled the oil companies and addressed the desires of OPEC members for higher prices and participation in the oil business.[41] U.S. consumers would have stoutly opposed international negotiations to raise energy prices, and State Department officials had, of course, declined to draw a contract acceptable to Venezuela on the Orinoco Tar Belt. At least publicly, no official conceded what one congressional committee drew as the overriding lesson of the energy crisis; that is, "in a democratic society, important questions of policy with respect to a vital commodity like oil, the lifeblood of an industrial society, cannot be left to private companies acting in accord with private interests and a closed circle of government officials."[42] The Nixon administration's only response to the looming crisis was to lift on May 1, 1973, the volume restrictions on imported oil. This was not a policy; it was a signal to OPEC that the United States was vulnerable.

The efforts of OPEC redounded to the benefit of Venezuela. Though it had been during the 1960s the most militant member of the cartel on the questions of raising prices and limiting production, Venezuela did not play between 1970 and 1973 a prominent role in the organization. In part, this reflected Venezuela's declining position in the international oil business. In 1970, Venezuela relinquished its position as the world's leading exporter of oil, a distinction it had held since 1928, to Saudi Arabia and Iran. And its proven reserves were a fraction of those of Middle Eastern and North African oil producers. While fulfilling nationalist aspirations, the initiatives the Caldera administration pushed—new taxes, production goals, the Reversion Law—were within the context of "Libyan leapfrog." The nation was surprised by the rapid developments in the oil trade. As late as March 1973, Oil Minister Pérez de la Salvia, in reporting to legislators, would only say "that we find ourselves in the center of a transitional stage—and with that—on the threshold of a new petroleum era."[43] Venezuela may have been, however, the last nation that the multinational oil companies coerced. Between 1971 and 1972, the companies thought they had with Middle Eastern producers a five-year contract allowing for gradual tax increases. The production cutbacks that Venezuela suffered in 1971–1972 may have been designed to force Venezuela to sign a similar agreement.[44] In any case, Venezuela contributed to OPEC by helping organize and keep the cartel together during the difficult 1960s. As one U.S. journalist commented, Oil Minister Pérez Alfonzo had been "the unlikely father of Arab power."[45]

Venezuela initially reacted to the U.S. energy crunch in an ungracious, even recriminatory manner. President Caldera set the tone when he gloated that "if the United States didn't have the intelligence to offer us a hemisphere preference . . . for so long, it now makes little difference to us." During the embargo, Venezuela faithfully exported oil to the United States, proving, as the embassy in Washington archly observed, "the reliability of Venezuela's oil in times of emergencies." Another official recalled "going begging in vain to Washington every year for long-term import quotas for our oil so that we could properly make development plans." But these annual requests, coming when oil was abundant, "fell on deaf ears." Now, Caldera asserted, "it is impossible to hide the fact that we have a very precious commodity . . . the need for which is increasing daily at an astonishing rate in every country."[46]

Such shrillness characterized Venezuelan-U.S. relations for more than a year after the imposition of the oil embargo. The chief spokesman for Venezuela in the acrimonious debate was its vigorous new president, Carlos Andrés Pérez. In a surprising electoral reversal, Acción Democrática surged back into power in December 1973. AD's leader was a protege of Rómulo Betancourt and a veteran party worker, who had served as minister of interior in the Betancourt administration and as legislator in the Chamber of Deputies since 1959. Drawing upon campaign tactics popular in the United States, Pérez brought his "democracy with energy" theme to the people by walking over three thousand miles. His efforts won him 49 percent of the vote, a broad mandate in a multicandidate election.[47]

While the momentous developments in international oil had not been a major campaign issue, President-elect Pérez immediately indicated that Venezuela would seize the historic moment. Charging that the foreign oil companies had irresponsibly allowed the oil reserves to dwindle, Pérez announced that Venezuela would not wait until 1983 to restore sovereignty over its natural resources. President Caldera, in his last state of the nation address, supported Pérez by urging his successor to nationalize the steel and oil companies. Beyond confirming that Venezuela would nationalize the extractive industries during his administration, President Pérez, partially heeding the advice of Pérez Alfonzo, approved a 12 percent reduction in oil output in 1974. The move, which was less than Pérez Alfonzo's recommendation to slash production in half, would support OPEC's prices and extend the life of the nation's reserves.[48]

In addition to ordering new oil policies, which he labeled "exercising responsibility with prudent audacity," Pérez launched a fiery attack on the industrial world in general and the United States in particular. In his inaugural address, he declared that Venezuela would "take up the defense of Latin American rights, trampled by the economic totalitarianism of the developed countries." His ambassador to Washington accused the United States of allowing "the existence of a vacuum in understanding with its hemispheric neighbors" and taking it "for granted that there will always be time for a meeting of the minds for us." Pérez insisted that the United States must repeal the economic sanctions against Cuba, relinquish the Panama Canal, and show more interest "in Venezuela and Latin America than in our raw materials." And, when the new oil prices were criticized for precipitating the world recession of 1974 and placing an onerous burden on the poorest nations, the Venezuelan reacted angrily. In an open letter to President Gerald R. Ford, he replied to the president's statement at the United Nations that OPEC was fomenting a global "economic confrontation." According to Pérez, this confrontation "has been created by the major powers, which refuse to allow the developing countries equal participation in the search for an indispensable balance in the terms of trade." In another open letter, to the World Food Conference in Rome, Pérez claimed that developed countries caused world famine and hunger by perpetuating "a wasteful economic system founded on unjust prices for the raw materials produced by the poor or impoverished countries." He jeered that, instead of dealing with the root causes of the crisis, "those who are primarily responsible for the situation will continue to salve their consciences with charity and philanthropy."[49]

The United States exchanged insults with Venezuela. Neither the Nixon nor the Ford administration thanked the Latin nation for increasing its exports of oil to the United States during the Arab embargo. The Nixon government also clumsily renewed its offer to help develop the Orinoco Tar Belt by arrogantly suggesting that "the further oil prices rise, the harder they will fall."[50] Congress added to the ill-will by excluding OPEC members from the "generalized system of preferences" attached to the Trade Act it passed in 1974. The new system, which was a response to Third World demands for expanded trade, gave preferential tariff treatment to designated exports of developing nations. In approving the Trade Act, Congress dismissed administration arguments that it should not punish oil producers, like Ecuador, Nigeria, Indonesia, and Venezuela, who had not participated in the embargo.[51] Although he reluctantly signed the Trade Act, President Ford, like Congress, unstintingly and indiscriminately criticized OPEC through 1974.[52]

This mutual recrimination, which had not been heard since the Cipriano Castro era, was short-lived. Led by Secretary of State Henry Kissinger, who perhaps belatedly recognized that global politics did not always revolve around military power, the United States offered to engage in a "new dialogue" with Latin America. While short on concrete proposals, Kissinger suggested discussions on such issues as commodity agreements, transfers of technology, the role of the multinational corporation, and new rules of foreign investment. The secretary hastened to add, however, that while prepared to listen attentively "we are convinced that the present economic system has generally served the world well" and that "history" has proven "poorer nations benefit most from an expanding world economy." Beyond listening to the economic grievances of Latin America, the Ford administration addressed two political concerns. It did not oppose a Venezuelan initiative to lift economic sanctions against Cuba at a meeting of the Organization of American States in November 1974, and it reopened negotiations with Panama on the future of the canal.[53]

Faced with some stark political realities, President Pérez also spoke and acted more temperately than he had during the first year of his rule. Pérez' critique of the international economic system implied a radical redistribution of global income and resources. This seemed a curious stance for a politician who as minister of interior jailed Venezuelan radicals and who represented a party and political system that favored gradual reform over dramatic change. Indeed, domestic concerns increasingly impinged upon the Venezuelan leader. The flush of oil income raised expectations in the *ranchos* and countryside that were difficult to satisfy immediately. The new money also offered numerous opportunities for wasteful spending and corruption. By mid-1975, an exasperated Pérez reportedly muttered that he could not "turn his back without some new scandal."[54]

The currents of world affairs also sobered Pérez. Diplomacy proved more complex than simply reciting Third World demands. Venezuela's representative to the Organization of American States embarrassed the nation, for example, by failing in late 1974 to line up enough votes to lift economic sanctions against Cuba. In addition, Venezuelans began to question the foreign policies of hemispheric neighbors. The nation grew wary of Brazil's expanding military and economic might, and it objected to Cuba's military adventures in Angola and Ethiopia.[55] On the other hand, President Pérez welcomed Secretary Kissinger's proclaimed new attitude toward Latin America. By 1975, he was conceding that, while global economic relationships needed to be changed permanently, the production and price of oil must take into account the requirements of industrial countries. Furthermore, he emphasized that good relations with the United States were "a matter of prime importance." Finally, Pérez' assessment of the warnings to temper his rhetoric that he reportedly received in stiff notes from Ford and Kissinger perhaps convinced him to be conciliatory.[56]

Both the moderate character of the Venezuelan political milieu and the limitations on the nation's political and economic independence were also revealed by the path the Pérez administration followed in nationalizing the foreign steel and oil properties. Compared to the turmoil that surrounded the oil law of 1943 and Pérez Alfonzo's policies in the early 1960s, the nationalization of iron-ore deposits on January 1, 1975, and of oilfields exactly one year later was remarkably orderly and peaceful. The government calculated the net book value of the foreigners' unamortized investments and then paid the steel companies approximately $100,000 and the oil companies $1.028 billion in interest-bearing Venezuelan bonds. The Pérez administration rejected the demand of socialists in Congress to expropriate without compensation. Moreover, it tempered nationalist fervor by allowing in the nationalization decree for a service-contract role for foreign companies and mixed government-industry enterprises. As President Pérez explained in July 1975 in a nationwide broadcast, "we believe that their technological capacity, their marketing means in the world, confer on these companies, as well as other public and private international entities, roles that may be taken advantage of in future negotiations with state entities." The government therefore signed technical assistance contracts with both the oil and the steel companies and agreed to sell the bulk of its oil output to the U.S. and British-Dutch enterprises.[57] And, when the nation decided to tap the Orinoco Tar Belt, it could, through article 5 of the nationalization law, offer foreign companies a partnership.

In "taking advantage" of the foreign companies' expertise, Venezuela was ruefully admitting that it had not escaped economic dependence upon the industrial world, particularly the United States. When Castro and Gómez invited the U.S. and British-Dutch oil companies to explore Venezuela, the impoverished nation lacked capital, technology, managerial talent, industrial skills, and, if oil was discovered, developed markets. The nation gradually rectified some of these deficiencies. After 1943 and especially after the overthrow of Pérez Jiménez, governments insisted that foreign companies train Venezuelans in both skilled and managerial positions. By 1974, Bethlehem Steel and U.S. Steel employed 3,500 people in Venezuela, but only 47 of them were from the United States. The government also encouraged students to study science and mathematics abroad. General Rafael Alfonzo Ravard, the first head of Petroven, the state's petroleum holding company, earned a degree in engineering, for example, from the Massachusetts Institute of Technology. The revolution in oil prices brought about, however, the most significant change in Venezuela's relations with foreign corporations, for, by giving the country capital, it took away the foreigners' power to counter any new regulation or tax proposal by threatening not to invest to expand oil and iron-ore reserves. But the country's technical, scientific, and entrepreneurial advances were slow and uneven. It continued to lack the educational and scientific base that could produce technological innovations, such as enhanced recovery methods in oil production. And operating the new state companies would drain the pool of managerial talent; the country would not be capable of also marketing and distributing all its petroleum products.

The response of the oil and steel companies and the State Department to nationalization underscored Venezuela's moderation. To be sure, the companies opposed nationalization and grumbled about their compensation. Claiming that their properties were worth at least $5 billion, oilmen waited until the last day before accepting the government's offer. If they had refused, the government would have expropriated their holdings and sent the matter of compensation to the Venezuelan Supreme Court. In view of the global energy crunch and the militancy of oil producers in the Middle East and North Africa, oil executives had little choice but to accept the one billion dollar offer.[58] But, as in 1943 when confronted with inevitable change, the oil companies again demonstrated that they could profit handsomely under the new rules. The companies knew the technical and entrepreneurial shortcomings of their host. As early as February 1974, the position of Creole Petroleum was that "even if early reversion does take place, Creole is optimistic that new long-term relationships may evolve between Venezuela and the corporation, which would allow the corporation to continue to provide important services to the Venezuelan nation on a mutually satisfactory and profitable basis." Less diplomatically, one cocksure executive told a journalist that "the Venezuelans can get the technicians but not even Moscow will share their technology with them." The marketing and technical assistance agreements that the companies won bore out their optimism and led one financial analyst to predict that "Exxon will make more money on Venezuelan oil than before nationalization."[59]

The Department of State deemed the transfer of private U.S. properties to Venezuela as satisfactory. The position of the United States, as outlined by Secretary of State Kissinger, was that, "while we do not recommend expropriation, and indeed, while it runs counter to the investment of private capital, which may be one of the best sources for the underdeveloped countries of capital, we do not, as a government, object to it if there is fair compensation and due legal process."[60] Assured by President Pérez that Venezuela would continue to sell oil to the United States, the department did not enter the government-industry negotiations, although Ambassador McClintock reportedly urged Washington to assist the companies. Oil executives flatly rejected any help, however, perhaps fearing that bilateral talks would lead to a marketing arrangement between the two countries. Such a development, the chairman of the board of Exxon warned stockholders, would be "a very serious challenge to the energy companies."[61] A month after the nationalization of the oil companies, Secretary Kissinger sanctioned the process by visiting Caracas. While there Kissinger applauded the nation's "new prosperity and power," while noting that U.S. private enterprise had contributed to this development. The secretary also thanked Venezuela for being an "important and reliable supplier of energy to the United States" and urged it not to seek confrontations with the industrial world.[62]

As Henry Kissinger indicated, the surge of oil income had transformed Venezuela, providing it with resources to pursue new economic development programs and to influence world affairs. And, as the secretary of state hoped, the Caribbean nation seemed inclined to use its new power and prosperity in a manner acceptable to the United States. The quadrupling of oil prices in 1973 allowed Venezuela's per capita income to rise from about $1,250 to approximately $2,300 in 1975, a figure nearly as high as the per capita incomes of European countries like Ireland and Greece. Far surpassing Argentina as Latin America's wealthiest nation, Venezuela had in early 1976 monetary reserves of $8.8 billion, which were equal to the reserves of all other Latin American nations combined. Moreover, the nation's financial security seemed assured for the rest of the twentieth century. By cutting oil production more than 40 percent from its 1970 high to approximately 2.2 million barrels a day, the government lengthened the life of the country's traditional oilfields. Nevertheless, Venezuela would collect more money from the sale of its oil between 1976 and 1980, for example, than it had between 1921 and 1973.[63]

This undreamed of affluence posed both opportunities and problems for the nation. Venezuela would have the income to emerge from the ranks of the poor nations and to provide its citizens with a generous standard of living. President Pérez decreed in 1974, for example, wage increases as high as 25 percent for the lowest-paid workers. Moreover, he accelerated the government's plans to construct a petrochemical complex, El Tablazo, in the west and a metallurgical complex near the iron ore resources in the east at Ciudad Guayana. These ambitious projects, reminiscent of other "*sembrar el petróleo*" campaigns, were designed both to diversify the economy and to induce people to leave teeming Caracas for the new jobs, a so-called poles of development strategy.[64] Yet, while these and other development projects seemed soundly conceived, it would be difficult for the state to reach the soaring expectations of the people. Venezuela's economic and social problems were serious. In the mid-1970s, at least one-third of the population lived poorly and probably one-fifth of the labor force was unemployed or underemployed. Life and work in the countryside remained dismal; agricultural workers, who comprised 20 percent of the population, earned a meager 2 percent of the national income. And, while it was one of the least densely settled countries in the Western Hemisphere, Venezuela too suffered from the pangs of the global population explosion, with one-half of its twelve million people under the age of fifteen.[65] Delivering services to and creating enough work for these young Venezuelans would surely test the efficiency of the nation's public and private sectors. The future would also witness whether Venezuela had the will to work for a socially just society. Prominent Venezuelans, such as Rómulo Betancourt, warned that the country would be gripped in a "crazy-collective hysteria" of extravagant spending that would perpetuate the gross inequities in the society while gravely distorting "our system of values."[66]

However Venezuela dispensed its wealth, U.S. businessmen felt confident that they would garner a substantial share of it. The effect of the nationalizations had been, of course, to reduce the economic influence of the United States in Venezuela. Moreover, Venezuela had altered its rules on trade and investment. In mid-1972, in a near final effort to pressure Washington into lifting the quotas on imported oil, the Caldera government had terminated the three-decade-old reciprocal trade agreement with the United States. Venezuela then joined two regional economic organizations, the Andean Pact and Sistema Económico Latinamericano (SELA). It also accepted Decision 24 of the Andean Pact, which limited profit remittances and foreign investment.[67] Yet, though wanting to widen its economic contacts, Venezuela found that it could not easily turn away from its traditional trading partner. Only industrial nations like the United States had the sophisticated technology, such as computers and telecommunications systems, that the nation wanted for its development programs. Furthermore, Venezuelan consumers had developed through the twentieth century tastes for U.S. products. As one bemused journalist observed, Venezuelans saw the United States both as "a Yankee exploiter and as supplier of fancy cars, rock music, and appliances that lasted longer than local ones."[68] As such, U.S. sales to Venezuela kept pace with the South American nation's new spending habits. The value of U.S. exports to Venezuela leaped from $924 million in 1972 to $2.62 billion in 1976. Eager to recycle "petrodollars," both the State and Commerce departments assisted trade expansion by preparing market analyses for exporters and assigning a commercial task force to the embassy in Caracas. What one commerce official dubbed "one of the most unique markets in the world" continued to be among the nation's ten best customers in the world and the second best, after Mexico, in Latin America.[69]

U.S. investors also found a newly rich Venezula hospitable. In 1972, the value of direct investment in Venezuela was $2.68 billion, with approximately $2 billion of it in the extractive industries. After nationalization, the value of U.S. investments recovered to $1.5 billion by the end of 1976. This new money was concentrated in manufacturing. As reflected also in the patterns of trade, Venezuela's dilemma was that economic modernization and independence seemed obtainable only through increased association with the multinational corporation. The Pérez administration discovered that, while it could purchase hardware, the nation lacked enough technicians and administrators to guide development. The government hoped that, by encouraging joint ventures between domestic and foreign capitalists, it would reap the benefits of foreign investment, like skilled personnel, without, as in the past, becoming subservient to it.[70]

In foreign affairs, Venezuela acted in consideration of the interests of the United States. It did not oppose the United States on any key East-West issue, and it continued to purchase its military equipment from the United States and its allies. Moreover, it spoke against using oil as a diplomatic club, and it readily sold its output to the United States. But, because of production cutbacks and the rapidly growing oil needs of the United States, Venezuela supplied in 1976 only about 13 percent of its northern neighbor's import needs.[71] In any case, Venezuela contributed many of these dollars, as the United States urged all OPEC members to do, to foreign aid. During the 1970s, Venezuela went from an aid recipient, as small programs like Food for Peace and the Peace Corps were ended, to among the world's most generous dispensers of aid. Wanting to prove the sincerity of its commitment to the Third World and also to invest some of its petrodollars abroad to prevent ruinous inflation at home, Venezuela gave in 1975, for example, 12 percent of its gross domestic product to foreign aid. It also loaned money to Central American and Caribbean nations for oil purchases.[72] This new role ironically left the South American nation open to charges of "imperialism." As an economist from the Dominican Republic noted, "1974 probably represents the close of a period that began in 1961 of great dependence of our country on the United States, and unfortunately the beginning of another period of economic dependence on Venezuela and other nearby oil producers."[73] Washington, however, applauded Venezuela for recognizing its "responsibility" to use its "surplus wealth to meet the human needs of the world's people."[74]

When, on January 1, 1976, President Pérez raised the Venezuelan flag over the nation's first own oil well, he told citizens that nationalization "is an act of faith in Venezuela and Venezuelans and in our capacity to assume and build our own destiny."[75] To observers of contemporary Venezuela, that destiny was not to effect a wholesale change in the relations between the industrial world and the developing nations. Instead, Venezuela seemed bound to use, in the words of one scholar, "its new international strength to pursue a reformist brand of nationalism aimed at reducing its dependence upon other capitalist states, and building its independence within their system." Both the nature of the nation's development, trade, aid, and investment policies and the character of its electorate, which opinion surveys found to be politically moderate, anticommunist, and wedded to developmentalist programs, supported that conclusion.[76]

Appendix. Statistical Data

Table A. U.S. Trade with Latin America and the World, 1910–1976 (Millions of Dollars)

	1910–1914[a]		1929		1933	
	Exports	*Imports*	*Exports*	*Imports*	*Exports*	*Imports*
Latin America	302	435	937	1106	240	329
Argentina	47	33	210	117	37	34
Brazil	32	111	109	208	30	83
Chile	14	23	56	102	5	12
Colombia	6	12	49	104	15	48
Cuba	63	122	129	207	25	58
Mexico	53	71	134	118	38	31
Venezuela	5	9	45	51	13	14
World	2,166	1,689	5,241	4,400	1,675	1,450

	1939		1945		1948	
	Exports	*Imports*	*Exports*	*Imports*	*Exports*	*Imports*
Latin America	549	518	1,261	1,637	3,166	2,352
Argentina	71	62	39	169	381	180
Brazil	80	107	219	311	497	514
Chile	27	41	52	135	105	179
Colombia	51	49	88	103	197	236
Cuba	82	105	196	338	441	375
Mexico	83	56	307	231	522	246
Venezuela	62	24	136	85	517	271
World	3,177	2,320	9,806	4,159	12,653	7,124

Table A (*continued*)

	1950		*1957*		*1960*	
	Exports	*Imports*	*Exports*	*Imports*	*Exports*	*Imports*
Latin America	2,671	2,910	4,567	3,769	3,455	3,529
Argentina	145	206	284	129	349	99
Brazil	353	715	484	700	426	570
Chile	72	160	195	196	195	193
Colombia	233	313	242	384	246	300
Cuba	460	406	618	482	223	357
Mexico	518	315	903	430	807	443
Venezuela	399	324	1,053	900	551	948
World	10,275	8,852	20,850	12,982	20,500	14,654

	1970		*1976*	
	Exports	*Imports*	*Exports*	*Imports*
Latin America	5,696	4,779	15,492	13,227
Argentina	441	172	544	308
Brazil	841	670	2,809	1,737
Chile	300	157	508	222
Colombia	395	269	703	655
Cuba	—	—	—	—
Mexico	1,704	1,219	4,990	3,598
Venezuela	759	1,082	2,678	3,574
World	43,226	39,963	114,997	120,667

[a]Average annual value of trade for five years (1910–1914).

Sources: For 1910–1914, U.S. Department of Commerce, Bureau of Foreign and Domestic Commerce, *Commerce Yearbook, 1922*, p. 505; for 1929, idem, *Commerce Yearbook, 1930*, I, 126; for 1933–1960, U.S. Department of Commerce, Office of Business Economics, *Business Statistics, 1961*, pp. 106–113; for 1970–1976, U.S. Department of Commerce, Bureau of Economic Analysis, *Survey of Current Business*, various issues.

Table B. U.S. Direct Investment in Latin America and the World, 1913–1976 (Millions of Dollars)

	1913[a]	*1929*	*1943*	*1950*	*1960*	*1970*	*1976*
Latin America	1,242	3,462	2,720	4,445	8,365	12,252	17,125
Argentina	40	332	380	356	472	1,281	1,366
Brazil	50	194	233	644	953	1,847	5,416
Chile	15	423	328	540	738	748	179
Colombia	2	124	117	193	424	698	654
Cuba	220	919	526	642	956	0	0
Mexico	800	682	286	415	795	1,786	2,976
Panama	5	29	110	58	405	1,251	1,961
Peru	35	124	71	145	446	688	1,364
Venezuela	3	233	373	993	2,569	2,704	1,506
Other Central America	36	206	173	245	319	624	680
Other	36	197	124	214	288	626	1,024
World	2,652	7,528	7,867	11,778	32,744	78,176	136,396

[a]Figures cited for 1913 include a small amount of portfolio investments.

Sources: For 1913, Winkler, *United States Capital in Latin America*, p. 275, and Lewis, *America's Stake*, p. 606; for 1929–1970, Subcommittee on Multinational Corporations of the Committee on Foreign Relations, *Multinational Corporations in Brazil and Mexico*, pp. 34–37; for 1976, U.S. Department of Commerce, Bureau of Economic Analysis, *Survey of Current Business* 58 (August 1978): 27.

Table C. World Production of Crude Petroleum, 1918–1976 (Millions of Barrels)

Year	*Iran*	*Mexico*	*Saudi Arabia*	*U.S.S.R.*	*U.S.A.*	*Venezuela*	*Total*
1918	8	64	—	27	356	—	504
1919	10	87	—	32	378	—	556
1920	12	157	—	25	443	—	689
1921	17	193	—	29	472	1	766
1922	22	182	—	36	558	2	859
1923	25	150	—	39	732	4	1,016
1924	32	140	—	45	714	9	1,014
1925	35	116	—	52	764	20	1,069
1926	36	90	—	64	771	37	1,097
1927	40	64	—	77	901	63	1,263
1928	43	50	—	85	901	106	1,325
1929	42	45	—	100	1,007	137	1,486
1930	46	40	—	126	898	137	1,410
1931	44	33	—	163	851	117	1,374
1932	49	33	—	156	785	117	1,311
1933	54	34	—	155	906	118	1,442
1934	58	38	—	174	908	136	1,522
1935	57	40	—	182	997	149	1,655
1936	63	41	—	186	1,100	155	1,792
1937	78	47	—	193	1,279	186	2,039
1938	78	38	—	205	1,214	188	1,988
1939	78	43	4	213	1,265	206	2,085
1940	66	44	5	219	1,353	186	2,150
1941	51	42	4	238	1,402	228	2,223
1942	72	35	5	227	1,387	148	2,080
1943	75	35	5	201	1,506	178	2,257
1944	102	38	8	275	1,678	257	2,592
1945	131	44	21	149	1,714	323	2,595
1946	147	49	60	158	1,734	388	2,745
1947	155	56	90	187	1,857	435	3,022

Table C (*continued*)

Year	*Iran*	*Mexico*	*Saudi Arabia*	*U.S.S.R.*	*U.S.A.*	*Venezuela*	*Total*
1948	190	59	143	218	2,020	490	3,433
1949	205	61	174	238	1,842	482	3,404
1950	242	72	200	266	1,974	547	3,803
1951	128	77	278	285	2,248	622	4,283
1952	8	77	302	322	2,290	660	4,531
1953	9	72	308	363	2,357	644	4,798
1954	22	84	348	427	2,315	692	5,017
1955	121	89	352	510	2,484	787	5,626
1956	197	91	361	612	2,617	899	6,125
1957	263	88	362	718	2,617	1,014	6,451
1958	301	94	370	826	2,449	951	6,608
1959	345	96	400	946	2,575	1,011	7,133
1960	386	99	456	1,079	2,575	1,042	7,674
1961	432	107	508	1,212	2,622	1,066	8,184
1962	482	112	555	1,360	2,677	1,168	8,882
1963	538	115	595	1,504	2,753	1,186	9,539
1964	619	116	628	1,644	2,787	1,242	10,311
1965	688	118	739	1,786	2,849	1,268	11,057
1966	771	121	873	1,948	3,028	1,230	12,016
1967	948	133	948	2,100	3,216	1,293	12,889
1968	1,039	142	1,114	2,272	3,329	1,319	14,093
1969	1,232	150	1,174	2,413	3,372	1,311	15,214
1970	1,397	157	1,387	2,595	3,517	1,353	16,711
1971	1,662	177	1,746	2,778	3,454	1,294	17,663
1972	1,839	185	2,202	2,896	3,455	1,178	18,601
1973	2,139	191	2,773	3,094	3,361	1,229	20,368
1974	2,198	238	3,096	3,374	3,203	1,086	20,538
1975	1,953	294	2,583	3,609	3,052	856	19,498
1976	2,168	327	3,140	3,822	2,972	839	21,187

Source: U.S. Department of Interior, Bureau of Mines, *Minerals Yearbook*, 1932–1976.

Table D. Mineral Oils, Crude and Refined, Imported into the U.S., 1946–1976 (Millions of Barrels)

Year	*Africa*	*Canada*	*Mexico*	*Middle East*	*Netherlands[a] Antilles*	*Venezuela*	*Total*
1946	—	—	6	—	47	73	138
1947[b]	—	—	11	3	54	80	159
1948	—	—	6	23	56	94	188
1949	—	—	9	37	73	106	239
1950	—	—	18	43	99	130	309
1951	—	—	17	39	97	131	308
1952	—	—	11	58	104	152	351
1953	—	3	12	80	91	161	383
1954	—	3	20	79	87	175	392
1955	—	17	23	102	99	208	470
1956	—	46	22	106	101	243	544
1957	—	57	14	88	102	292	591
1958	—	32	12	131	127	295	652
1959	—	36	13	128	123	318	674
1960	—	44	6	120	117	334	687
1961	2	69	15	125	116	321	717
1962	10	91	18	115	109	331	760
1963	9	97	17	112	113	329	775
1964	19	109	17	115	122	342	827
1965	26	118	17	131	132	363	901
1966	32	140	17	117	126	372	939
1967	23	164	18	76	131	342	926
1968	58	185	16	80	144	324	1,039
1969	84	222	16	70	164	319	1,156
1970	46	280	15	70	175	361	1,248
1971	74	313	10	139	158	372	1,433
1972	183	406	8	173	155	351	1,735
1973	308	484	6	311	213	414	2,283
1974	368	390	3	380	187	357	2,231
1975	508	309	26	426	121	256	2,210
1976	738	219	32	687	101	256	2,670

[a]Virtually all oil imported into the United States from the Netherlands Antilles, the islands of Aruba and Curaçao, was produced in Venezuela. The oil companies refined some of their Venezuelan production in Aruba and Curaçao, before shipping it to the United States.

[b]After 1947, the United States always imported more petroleum than it exported. As a percentage of domestic demand, net imports (total imports minus total exports) rose from 2.4 percent in 1948 to 16 percent in 1958 to 20 percent in 1968 to 36 percent in 1973 to 41 percent in 1976.

Source: Bureau of Mines, *Minerals Yearbook*, 1947–1976.

Notes

1. The Caribbean Sphere of Influence

1. Miguel Izard, *Series estadísticas para la historia de Venezuela*, p. 10; Edwin Lieuwen, *Venezuela*, p. 10.
2. Lieuwen, *Venezuela*, pp. 14–16, 117.
3. Ibid., pp. 122–123; Miguel Izard, "Período de la independencia y la Gran Colombia," in Miguel Izard (ed.), *Política y economía en Venezuela, 1810–1976*, pp. 21–30; Benjamin A. Frankel, "La Guerra Federal y sus secuelas," in Izard (ed.), *Política*, pp. 153–161; Robert L. Gilmore, *Caudillism and Militarism in Venezuela, 1810–1910*, p. 13; William Maurice Sullivan, "The Rise of Despotism in Venezuela: Cipriano Castro, 1899–1908" (Ph.D. dissertation), pp. 9–15.
4. Lieuwen, *Venezuela*, pp. 120–121; Sullivan, "Despotism in Venezuela," pp. 26–29; George Edmund Carl, "British Commercial Interest in Venezuela during the Nineteenth Century (Ph.D. dissertation), pp. 16, 151–152.
5. Edwin Lieuwen, *Petroleum in Venezuela: A History*, pp. 4–6; Ralph Arnold, et al., *The First Big Oil Hunt: Venezuela, 1911–1916*, p. 89; Frederick C. Gerretson, *History of the Royal Dutch*, IV, 273.
6. William H. Gray, "American Diplomacy in Venezuela, 1835–1860," *Hispanic American Historical Review* 20 (November 1940): 551–574; P. F. Fenton, "Diplomatic Relations of the United States and Venezuela, 1880–1915," *Hispanic American Historical Review* 8 (August 1928): 330; Benjamin Adam Frankel, "Venezuela and the United States, 1810–1888" (Ph.D. dissertation).
7. Miriam Hood, *Gunboat Diplomacy, 1895–1905: Great Power Pressure in Venezuela*, pp. 134, 168; Carl, "British Commercial Interest," pp. 52, 143–145.
8. Carl, "British Commercial Interest," pp. 44–65, 135–145; Frankel, "Venezuela and the United States," p. 215; Izard, *Series estadísticas*, pp. 202–203.
9. Charles S. Campbell, *The Transformation of American Foreign Relations, 1865–1900*, pp. 194–221; Frankel, "Venezuela and the United States," pp. 407–416.

10. Walter LaFeber, *The New Empire: An Interpretation of American Expansionism, 1860–1898*, pp. 269–278.
11. Ibid., pp. 251–253, 281; Albert K. Steigerwalt, *The National Association of Manufacturers, 1895–1914*, pp. 50–51; Charles Edward Carreras, "United States Economic Penetration of Venezuela and Its Effects on Diplomacy, 1895–1906," (Ph.D. dissertation), pp. 153–192.
12. U.S. Department of State, *Papers Relating to the Foreign Relations of the United States, 1895*, I, 558 (hereafter cited as *FRUS* followed by the appropriate year).
13. Winfield Burggraaff, "Venezuelan Regionalism and the Rise of Táchira," *The Americas* 25 (October 1968): 162–169; Domingo Alberto Rangel, *Los andinos en el poder*, p. 189; Sullivan, "Despotism in Venezuela," pp. 87–88.
14. Roosevelt quoted in Howard K. Beale, *Theodore Roosevelt and the Rise of America to World Power*, p. 406; Sullivan, "Despotism in Venezuela," pp. 183–188.
15. Beale, *Roosevelt*, pp. 395–431; *FRUS, 1904*, pp. 788–803; unsigned memorandum, May 30, 1902, Microfilm Reel #27, Series 1, Papers of Theodore Roosevelt, Library of Congress, Washington D.C.; Warren G. Kneer, *Great Britain and the Caribbean, 1901–1913: A Study in Anglo-American Relations*, pp. 32–63.
16. *FRUS, 1904*, pp. xli–xlii.
17. Mariano Picón-Salas, *Los días de Cipriano Castro*, pp. 236–239, 263–265; Sullivan, "Despotism in Venezuela," pp. 419–455.
18. Oray E. Thurber, *The Venezuelan Question: Castro and the Asphalt Trust*; John C. Rayburn, "United States Investments in Venezuelan Asphalt," *Inter-American Economic Affairs* 7 (Summer 1953): 34; Sullivan, "Despotism in Venezuela," pp. 302–303, 374; *FRUS, 1905*, pp. 919–1002.
19. Gretchen Kreuter, "The Orinoco Company and American-Venezuelan Relations, 1895–1911" (paper), pp. 19–20.
20. Beale, *Roosevelt*, pp. 405–406 (n. "a"; Sullivan, "Despotism in Venezuela," pp. 476–509.
21. W. Stull Holt, *Treaties Defeated by the Senate*, pp. 219–220.
22. Philip C. Jessup, *Elihu Root*, I, 493–499; Elting Morison (ed.), *The Letters of Theodore Roosevelt*, IV, 917; Embert J. Hendrickson, "Root's Watchful Waiting and the Venezuelan Controversy," *The Americas* 23 (October 1966): 120. For U.S. claims against Venezuela, see *FRUS, 1905*, pp. 919–1002, and U.S. Congress, Senate, Committee on Foreign Relations, *Correspondence Relating to Wrongs Done to American Citizens by the Government of Venezuela*, Sen. Doc. 413, 60th Cong., 1st sess. For rupture of relations, see *FRUS, 1908*, pp. 774–796, 820–821.
23. Sullivan, "Despotism in Venezuela," pp. 600–624.
24. *FRUS, 1909*, p. 609.
25. Ibid., pp. 609–612, 617–622; Harold F. Peterson, *Diplomat of the Americas: A Biography of William I. Buchanan*, pp. 326–341; Ministerio de

Relaciones Exteriores, *El libro amarillo de los Estados Unidos de Venezuela, 1909,* pp. xix–xxi, 103–171 (hereafter cited as *Libro amarillo* with appropriate year).

26. John Sleeper, chargé, to State Department, January 5, 1907, File 3136/6, General Records of the Department of State, Record Group 59, National Archives, Washington D.C. (hereafter cited as DSR with appropriate file and document number); John Brewer, legation archivist, to State Department, November 29, 1908, DSR 3136/69–71; Roosevelt quoted in Peterson, *Buchanan,* p. 326; Hendrickson, "Watchful Waiting," pp. 124–127; "A Nicely Timed Revolution," *Nation,* December 31, 1908, p. 645.
27. *FRUS, 1909,* p. 609; Brewer, archivist, to State Department, December 27, 1908, DSR 3136/93–96; Peterson, *Buchanan,* p. 338.
28. J. Fred Rippy and Clyde E. Hewitt, "Cipriano Castro, 'Man without a Country,'" *American Historical Review* 55 (October 1949): 36.
29. Ibid., pp. 40–43.
30. Ibid., pp. 45–52. The United States continued to keep Castro under scrutiny after he returned to Europe; see despatches between May 1911 and May 1912 in DSR 831.00/322–509.
31. Pedro Esequiel Rojas, minister of Venezuela, to Charles D. Hilles, secretary to the president, May 11, 1911, Series 6, File 463, Venezuela, Microfilm Reel #385, Papers of William Howard Taft, Library of Congress, Washington D.C.; *FRUS, 1912,* p. xxiv; Latin American Division memorandum, December 2, 1911, DSR 831.34/48; Sheldon Whitehouse, chargé, to State Department, April 28, 1910, DSR 831.00/296.
32. *FRUS, 1912,* pp. 1355–1364; Manuel Matos, minister of relations, to Taft, March 25, 1912, Series 6, File 463, Venezuela, Reel #385, Taft Papers.
33. U.S. Department of Commerce, Bureau of Foreign and Domestic Commerce, *Yearbook, 1922,* pp. 505–506; Steigerwalt, *National Association of Manufacturers,* p. 52; Kreuter, "Orinoco Company," p. 21; Max Winkler, *Investments of United States Capital in Latin America,* p. 275.
34. Gerretson, *Royal Dutch,* IV, 275; Lieuwen, *Petroleum,* pp. 12–31. A hectare equals 2.471 acres.
35. Arnold, *Oil Hunt,* p. 77; Gerretson, *Royal Dutch,* IV, 275–278; Michael O'Shaughnessy, *Venezuelan Oil Fields: Developments to September 1st, 1924,* pp. 18–19.
36. Gerretson, *Royal Dutch,* IV, 278–281; Lieuwen, *Petroleum,* pp. 14–15.
37. Lieuwen, *Petroleum,* pp. 10–12.
38. John A. DeNovo, "The Movement for an Aggressive American Oil Policy Abroad, 1918–1920," *American Historical Review* 61 (July 1956): 854–860.
39. William H. Libby of Standard Oil of New Jersey to State Department, March 19, 1914, DSR 831.6363/2; H. C. von Struve, consul in Caraçao, to State Department, March 24, 1915, DSR 831.6363/8; George S. Gibb and Evelyn H. Knowlton, *History of Standard Oil (New Jersey),* vol. II: *The Resurgent Years, 1911–1927,* p. 384.

40. Walter V. and Marie V. Scholes, *The Foreign Policies of the Taft Administration*, pp. 40–80.
41. Wilson's speech is in *Congressional Record*, 63rd Cong., 1st sess., Vol. 50, Part 6, pp. 5845–5846. See also "Declaration of Policy with Regard to Latin America," March 12, 1913, *FRUS,1913*, p. 7.
42. Karl M. Schmitt, *Mexico and the United States, 1821–1973: Conflict and Coexistence*, pp. 128–136.
43. Domingo Alberto Rangel, *Gómez: El amo del poder*, pp. 293–306; Daniel Joseph Clinton, *Gómez: Tyrant of the Andes*, pp. 126–129, 147–154; Sullivan, "Despotism in Venezuela," pp. 448–450, 619.
44. J. L. Salcedo-Bastardo, *Historia fundamental de Venezuela*, pp. 450–452; Consul Thomas Voetter, La Guaira, to State Department, September 14, 1913, DSR 831.00/577; Minister Henry Tennant to State Department, August 15, 1913, DSR 831.00/572, and September 27, 1913, DSR 831.00/601; Minister Preston McGoodwin to State Department, January 17, 1914, DSR 831.00/621, and February 2, 1914, DSR 831.00/627; Latin American Division memorandum by Rutherford Bingham, July 15, 1914, DSR 831.00/699.
45. Tennant to State Department, September 27, 1913, DSR 831.00/601; Latin American Division memorandum by Bingham, January 13, 1914, DSR 831.00/between 618 and 619.
46. McGoodwin to State Department, February 2, 1914, DSR 831.00/627, and March 30, 1914, DSR 831.00/643; Lansing to McGoodwin, April 24, 1914, DSR 831.00/643.
47. Memorandum by former Minister Tennant, Washington, [n.d.], DSR 831.00/between 618 and 619; McGoodwin to State Department, April 22, 1914, DSR 831.00/648.
48. Latin American Division memorandum [unsigned], November 19, 1913, 831.00/698.
49. Thomas A. Bailey, *The Policy of the United States toward the Neutrals, 1917–1918*, pp. 305–338; Percy A. Martin, *Latin America and the War*, pp. 461–473.
50. McGoodwin to State Department, April 23, 1917, DSR 711.31/104, June 7, 1917, DSR 831.00/794, June 19, 1917, DSR 862.20231/55, and October 29, 1918, DSR 831.00/864.
51. Lansing to McGoodwin, May 28, 1917, DSR 831.014/10; McGoodwin to State Department, June 24, 1917, DSR 831.014/13, August 26, 1917, DSR 831.00/798, and October 13, 1917, DSR 831.911/14.
52. Russell to Lansing, November 8, 1917, DSR 831.00/816; Stewart memorandum, January 5, 1918, DSR 831.00/833.
53. Lansing to Wilson, January 5, 1918, DSR 831.00/833–833a; Wilson to Lansing, February 16, 1918, DSR 831.00/834½.
54. Dana G. Munro, "Pan Americanism and the War," *North American Review* 208 (November 1918): 716. It should be noted, however, that Lord Reading, the British ambassador in Washington, reported that the United States inquired about British reaction to a revolution in Venezuela. The Foreign Office replied that it was "indifferent" to such an

eventuality; see despatches March 4 and March 6, 1918, Class 371, vol. 14457, doc. 40729, Records of the Foreign Office, Public Record Office, London (hereafter cited as FO with appropriate class, volume/document number).

55. Emil Sauer, consul in Maracaibo, to State Department, March 26, 1918, DSR 831.00/831. This despatch contains a plan for an invasion of Venezuela by anti-Gómez exiles.
56. F. Mayer, Latin American Division, to secretary of state, "Policy of the United States toward Venezuela," September 4, 1918, DSR 711.31/106.
57. Gómez to Wilson, July 1, 1919, Series 4, File 1224, Microfilm Reel #312, Papers of Woodrow Wilson, Library of Congress, Washington D.C.
58. Latin American Division memorandum by Mayer, February 1, 1918, DSR 831.00/824.

2. Open Door Diplomacy, 1919–1929

Portions of this chapter first appeared as "Anglo-American Rivalry for Venezuelan Oil, 1919–1929," *Mid-America* 58 (April–July 1976): 97–110. I gratefully acknowledge permission to use this material.

1. Quoted in DeNovo, "Aggressive American Oil Policy Abroad," pp. 854–856.
2. Ibid., pp. 856–857; Manning and Smith's opinions are in U.S. Congress, Senate, Special Subcommittee Investigating Petroleum Resources, *American Petroleum Interests in Foreign Countries*, 79th Cong., 1st sess., p. 229.
3. John A. DeNovo, *American Interests and Policies in the Middle East, 1900–1939*, pp. 176–202.
4. *FRUS, 1919*, I, 167.
5. Dwyre to State Department, April 12, 1920, DSR 831.6363/24; McGoodwin to State Department, November 11, 1919, DSR 831.6363/18, April 15, 1920, DSR 831.6363/25, and June 14, 1920, DSR 831.6363/39; Assistant Secretary of State Alvey A. Adee to McGoodwin, May 12, 1920, DSR 831.6363/25.
6. Dormer to Foreign Office, October 24, 1919, FO 371/A568/47.
7. U.S. Department of Commerce, Bureau of Foreign and Domestic Commerce, *American Direct Investments in Foreign Countries*, Trade Information Bulletin 731, p. 23; Julius Klein, "Economic Rivalries in Latin America," *Foreign Affairs*, December 15, 1924, p. 240; Cleona Lewis, *America's Stake in International Investments*, p. 606; J. Fred Rippy, *British Investments in Latin America, 1822–1949*, pp. 68, 75–76.
8. Comments of Second Secretary Thomas M. Snow on Ambassador to Chile John C. Vaughn's despatch to Foreign Office, September 1921, FO 371/A8333/436/51.
9. Lieuwen, *Petroleum*, pp. 15, 20–21; Ludwell Denny, *We Fight for Oil*, p. 111; McGoodwin to State Department, June 26, 1920, DSR 831.00/942.

10. Beaumont to Foreign Office, June 30, 1919, FO 371/7309/106603.
11. Lieuwen, *Petroleum*, p. 21. In the 1920s a bolívar (Bs.) was worth approximately twenty cents.
12. McGoodwin to State Department, November 25, 1919, DSR 831.00/919, and March 31, 1920, DSR 831.00/931.
13. C. K. McFadden to State Department, May 3, 1920, DSR 831.6363/22; Adee to McFadden, May 29, 1920, DSR 831.6363/27.
14. Comment of Weakley on Dormer despatch to Foreign Office, May 4, 1920, FO 371/A2895/2895/47.
15. *FRUS, 1921*, II, 934; unsigned State Department memorandum, September 9, 1920, DSR 831.6363/43; Dormer to Foreign Office, May 4, 1920, FO 371/A2895/2895/47, June 25, 1920, FO 371/A5152/1805/47, and July 23, 1920, FO 371/A5086/2895/47; Lieuwen, *Petroleum*, p. 22.
16. Dormer's diplomacy can be found in his despatches to the Foreign Office: March 26, 1920, FO 371/A2523/1764/47, June 14, 1920, FO 371/4784/2895/47, and November 8, 1920, FO 371/A8395/2895/47. See also D. E. Alves, president of British Oil Fields Company, to Foreign Office, June 7, 1920, FO 371/A3694/2895/47. British investors controlled 40 percent of Royal Dutch–Shell; Dutch investors controlled the other 60 percent. The Foreign Office, however, normally handled the company's diplomatic problems.
17. *FRUS, 1921*, II, 935; McGoodwin to State Department, March 25, 1921, DSR 831.6363/53.
18. John H. Murray to Sumner Welles, April 27, 1921, DSR 831.6363/93.
19. White to State Department, November 21, 1921, DSR 831.6363/94. After he learned of the department's disappointment, McGoodwin offered two other explanations for the revalidation of the Colon concession (McGoodwin to State Department, June 14, 1921, DSR 831.6363/62–63).
20. DeNovo, *Middle East*, pp. 176–202.
21. Geddes to Foreign Office, April 22, 1922, FO 371/A2715/2715/47. The U.S. consulate general in London, Robert Skinner, also predicted the concession transfer would improve Anglo-American relations (Skinner to State Department, May 2, 1922, DSR 831.6363/98).
22. Lieuwen, *Petroleum*, pp. 29–31; Rangel, *Gómez*, p. 300; Rómulo Betancourt, *Venezuela: Oil and Politics*, p. 32.
23. Craigie's comment on Henry Hobson despatch to Foreign Office, April 17, 1925, FO 371/A2612/1201/47.
24. Quotation is from letter from Ralph Arnold, representing the American Institute of Mining and Metallurgical Engineers, to Secretary of State Hughes, April 23, 1921, File: Oil 1920–1921, Box 452, Commerce Papers of Herbert Hoover, Hoover Presidential Library, West Branch, Iowa. For the legation's aid to U.S. oilmen, see, for examples, secretary of state to Caracas legation, January 25, 1922, DSR 831.6363/85b; secretary of state to Minister Willis C. Cook, October 27, 1923,

DSR 831.6363/142; Cook to State Department, July 20, 1928, DSR 831.6363/396, and August 13, 1928, DSR 831.6363/398.

25. Memorandum by Young, September 21, 1922, DSR 831.6375C23/35. In his perceptive study, *Informal Entente: The Private Structure of Cooperation in Anglo-American Economic Diplomacy, 1918–1928,* Michael J. Hogan argues that by 1922 the British had abandoned their attempts to establish a state-sponsored, preferential system of control over the world's oil resources. Instead, they had accepted a U.S. vision that "cooperation among private petroleum interests was a more acceptable means of regulating the development of oil resources than either wide-open competition or government intervention and management" (pp. 159–171). While persuasive in analyzing the goals of U.S. officials, such as Secretary of State Hughes and Secretary of Commerce Hoover, Hogan perhaps exaggerated the spirit and extent of Anglo-American cooperation in Venezuela. As indicated in this text, Foreign Office officials, recognizing their country's war-weakened condition, accepted some cooperation with U.S. oil companies because they could not prevent their intrusion into Venezuela. Moreover, they never relinquished the faint hope that the government-owned Anglo-Persian Oil Company might purchase concessions in Venezuela. See, for examples, Minister William Seeds to Foreign Office, January 12, 1926, FO 371/A641/641/47, and Sir John Cadman of Anglo-Persian Oil to Under Secretary Robert Vansittart, March 23, 1934, FO 371/A2508/2508/47. See also Emily S. Rosenberg's critique of Hogan in "Anglo-American Economic Rivalry in Brazil during World War I," *Diplomatic History* 2 (Spring 1978): 146 n. 51.
26. Leland Harrison to Cook, December 22, 1925, DSR 831.6363/293. Cook probably never spoke to Gómez, for Anglo-Persian officials decided, after the September visit, not to invest in Venezuela. They knew that Gómez would turn a "blind eye" toward their acquiring concessions, but they feared trouble after the dictator died (Seeds to Foreign Office, January 12, 1926, FO 371/A641/641/47).
27. The department refused in 1924–1925, for example, to mediate a dispute between U.S. companies over what facilities should be constructed to transport oil from Venezuela (Latin American Division memorandum by Young, February 6, 1925, DSR 831.156L33/33; Cook to State Department, April 18, 1925, DSR 831.156/38; Francis White to Young, May 12, 1925, DSR 831.156/46).
28. Special Subcommittee Investigating Petroleum Resources, *American Petroleum,* p. 323; Secretary of Interior Hubert Work to Hoover, July 7, 1927, File: Federal Oil Conservation Board—1927, Box 453, Commerce Papers of Hoover, Hoover Library.
29. Herbert Hoover, *The Memoirs of Herbert Hoover,* vol. II: *The Cabinet and Presidency,* p. 69; Joseph Brandes, *Herbert Hoover and Economic Diplomacy: Department of Commerce Policy, 1921–1933,* pp. 34–35, 134–135; Hoover to Hughes, January 9, 1923, DSR 831/6363/122.

30. Lieuwen, *Petroleum*, pp. 34–38; Rangel, *Gómez*, pp. 300–306, 395; Bennett to Foreign Office, April 28, 1924, FO 371/A3245/2967/47.
31. Cook to State Department, March 7, 1924, DSR 831.6363/160; Chargé Frederick Chabot to State Department, April 23, 1924, DSR 831.6363/176, and May 9, 1924, DSR 831.6363/196.
32. Secretary of state to Chabot, March 26, 1924, DSR 831.6363/160.
33. Chabot to State Department, April 5, 1924, DSR 831.6363/171. Chabot sent a formal note to the Foreign Ministry when Hughes wanted Chabot to determine first if there was discrimination and only then to speak to Venezuelan officials (Hughes to Chabot, April 25, 1924, DSR 831.6363/172).
34. Joseph S. Tulchin, *The Aftermath of War: World War I and U.S. Policy toward Latin America*, pp. 124–129.
35. Chabot to State Department, April 23, 1924, DSR 831.6363/176; Hughes to Chabot, April 26, 1924, DSR 831.6363/176.
36. Chabot to State Department, April 26, 1924, DSR 831.6363/180; Hughes to Chabot, May 3, 1924, DSR 81.6363.181.
37. Cook to State Department, January 9, 1925, DSR 831.6363/253; Harrison to Cook, February 10, 1925, DSR 831.6363/253; Kellogg to Cook, April 9, 1925, DSR 831.6363/264. The Foreign Office records are silent on this incident, although British diplomats kept the Foreign Office well informed about Venezuelan petroleum developments. It probably can be assumed that Shell never approached Compañía Venezolana and that Gómez repeated his Stinnes ruse to stimulate sales to U.S. oilmen.
38. Joan Hoff Wilson, *American Business and Foreign Policy, 1921–1933*, p. 9. Joseph Tulchin holds that the State Department's response to the Stinnes–Compañía Venezolana incident clearly demonstrates the U.S. commitment to the Open Door. Tulchin does not discuss the Anglo-Persian and Shell–Compañia Venezolana episodes. He also does not analyze U.S. policies toward Venezuela after 1924 (*Aftermath*, p. 154).
39. Special Subcommittee Investigating Petroleum Resources, *American Petroleum*, pp. 58–59; Michael O'Shaughnessy, *Venezuelan Oil Handbook*, p. 5.
40. DeNovo, *Middle East*, pp. 199–202; Hogan, *Informal Entente*, p. 217.
41. Lieuwen, *Petroleum*, pp. 38–39; *Oil and Gas Journal*, December 20, 1923, pp. R1–R23 (hereafter cited as *OGJ* with appropriate date and page).
42. Lieuwen, *Petroleum*, pp. 42–43; Paul H. Giddens, *Standard Oil Company (Indiana): Oil Pioneer of the Middle West*, pp. 240–247. Standard of Indiana acquired Lago Petroleum in 1925.
43. Federico G. Baptista, *Historia de la industria petrolera en Venezuela*, pp. 31–32; Lieuwen, *Petroleum*, p. 44.
44. Special Subcommittee Investigating Petroleum Resources, *American Petroleum*, p. 337; Franklin Tugwell, *The Politics of Oil in Venezuela*, p. 10.
45. Lieuwen, *Petroleum*, pp. 60–62; Rangel, *Gómez*, p. 394; Henrietta

M. Larson, Evelyn H. Knowlton, and Charles S. Popple, *History of Standard Oil Company (New Jersey)*, vol. III: *New Horizons, 1927–1950*, pp. 41–42; Cook to State Department, October 23, 1925, DSR 831.156L33/67; *OGJ*, September 25, 1924, p. 24.

46. U.S. Department of Commerce, Bureau of Foreign and Domestic Commerce, *Commerce Yearbook, 1930*, p. 126; Izard, *Series estadísticas*, pp. 202–203; Burton I. Kaufman, "United States Trade and Latin America: The Wilson Years," *Journal of American History* 58 (September 1971): 342–362; Rosenberg, "Economic Rivalry," pp. 151–152; Tulchin, *Aftermath*, pp. 38–40; Consul Alexander Sloan, Maracaibo, to State Department, November 23, 1925, DSR 711.62/1; Paul Daniels to State Department, June 30, 1938, DSR 611.3131/207.
47. Joseph Zettler and Frederick Cutler, "United States Direct Investments in Foreign Countries," *Survey of Current Business* 22 (December 1952): 8; Lewis, *America's Stake*, p. 588; Council on Foreign Relations, Walter Lippmann, *The United States in World Affairs, 1932*, pp. 51–52.
48. O'Shaughnessy, *Venezuelan Oil Handbook*, pp. 8–9; *OGJ*, December 29, 1927, p. 49; Lieuwen, *Venezuela*, pp. 24–29; Gibb and Knowlton, *Standard Oil: The Resurgent Years*, p. 388; Herbert Klein, "American Oil Companies in Latin America: The Bolivian Experience," *Inter-American Economic Affairs* 18 (Autumn 1964): 53–54; Tugwell, *Politics of Oil*, p. 17. For an account of one official who opposed the oil law of 1922, see Aníbal R. Martínez, *Gumersindo Torres*, pp. 85–128.
49. Denny, *We Fight*, p. 114; Rippy, *British Investments*, pp. 82, 123; Luis Vallenilla, *Oil: The Making of a New Economic Order*, pp. 40–41.
50. Cook to State Department, June 1, 1923, DSR 831.00/1182; Secretary of state to secretary of navy, June 13, 1923, DSR 831.00/1183; Cook to State Department, June 19, 1923, DSR 831.00/1185, and July 10, 1923, DSR 831.00/1190.
51. Clinton, *Gómez*, pp. 198–203.
52. Commander of Special Service Squadron J. H. Dayton to chief of naval operations, September 19, 1923, Secret and Confidential Correspondence File 117-20:2, Records of Special Service Squadron, General Records of the Department of Navy, Record Group 80, National Archives, Washington, D.C.
53. Cole to chief of naval operations, July 9, 1923, DSR 831.00/1192.
54. Dana G. Munro, *The United States and the Caribbean Republics, 1921–1933*, pp. 9–15; Tulchin, *Aftermath*, pp. 234–253; Sumner Welles, "Is America Imperialistic?" *Atlantic Monthly* 134 (September 1924): 413.
55. Keeling to Foreign Office, March 3, 1933, FO 371/A3901/3901/47.
56. Edward Gerald Duffy, "Politics of Expediency: Diplomatic Relations between the United States and Venezuela during the Juan Vicente Gómez Era" (Ph.D. dissertation), pp. 203–209; Chargé C. Van H. Engert to State Department, December 24, 1929, DSR 831.00/1450. British diplomats in Venezuela believed that Ministers McGoodwin and Cook ac-

cepted bribes from Gómez (Dormer to Foreign Office, January 1, 1921, FO 371/A618/618/47; Beaumont to Foreign Office, April 19, 1922, FO 371/A2631/221/47).

57. Gómez quoted in Chargé Warden McKee Wilson to State Department, January 7, 1932, DSR 710.g/47; Clinton, *Gómez*, pp. 139–140; Morris Gilbert, "A Dictator Who Is Now a Patriarch," *New York Times*, July 14, 1929, sec. 5, p. 6.
58. For example, Minister George Summerlin to State Department, December 23, 1929, DSR 831.00/1449.
59. Betancourt, *Venezuela*, p. 48; Rafael Gallegos Ortíz, *La historia política de Venezuela de Cipriano Castro a Pérez Jiménez*, pp. 154–155; Rangel, *Gómez*, p. 400.
60. Bennett to Foreign Office, April 28, 1924, FO 371/A3245/2967/47; White to State Department, November 14, 1921, DSR 831.00/1125. Some Venezuelans have defended Gómez' rule: Pedro Manuel Arcaya, *The Gómez Regime in Venezuela and Its Background*, pp. 205–226; Laureano Vallenilla Lanz, *Cesarismo democrático*, pp. 179–183; for a "balanced" interpretation of Gómez, see Pablo Emilio Fernández, *Gómez: El rehabilitador*, pp. 14–15.
61. Miguel Acosta Saignes, *Latifundio*, pp. 58, 109; Betancourt, *Venezuela*, pp. 65, 72; Gallegos Ortíz, *La historia política*, p. 125; Mariano Picón-Salas et al., *Venezuela independiente, 1810–1960*, p. 152; Rangel, *Los andinos*, p. 294; Winfield J. Burggraaff, *The Venezuelan Armed Forces in Politics, 1935–1959*, pp. 15–19.
62. Stabler to State Department, September 22, 1927, DSR 831.00/1337$^1/_2$.
63. Lieuwen, *Petroleum*, pp. 52–53; Sloan to State Department, July 30, 1925, DSR 831.504/9.
64. Daniel H. Levine, *Conflict and Political Change in Venezuela*, p. 17; John Duncan Powell, *Political Mobilization of the Venezuelan Peasant*, pp. 47–50; Sloan to State Department, April 30, 1926, DSR 831.5017/1.
65. Lieuwen, *Petroleum*, p. 52.
66. U.S. Department of Commerce, Bureau of Foreign and Domestic Commerce, C. J. Dean, *Commercial and Industrial Development of Venezuela*, Trade Information Bulletin 783, pp. 2–32.
67. Lieuwen, *Petroleum*, pp. 42–44; Sloan to State Department, June 30, 1926, DSR 831.401/original; *The Lamp* 9 (June 1926): 15–16. *The Lamp* is published by Standard Oil of New Jersey.
68. Anslinger to State Department, January 17, 1924, DSR 831.00/1221; William Seeds to Foreign Office, January 12, 1926, FO 371/A641/641/47.
69. Summerlin to State Department, August 24, 1934, DSR 831.00/1519.
70. Cook to State Department, October 23, 1925, DSR 831.504/11; Irvine L. Lenroot to Hoover, November 14, 1928, General Correspondence, Lenroot, Irvine File, Box 43, Pre-Presidential Papers, Hoover Library; see also letter of Francis Loomis, former minister to Venezuela, to Calvin Coolidge, December 18, 1928, Series 1, File 3085-Venezuela, Micro-

film Reel #174, Papers of Calvin Coolidge, Library of Congress, Washington D.C.; Kellogg to legation in Caracas, September 13, 1926, DSR 831.504/13a, and October 14, 1926, DSR 831.504/15; Cook to State Department, June 25, 1927, DSR 831.504/38.

71. Robert Neal Seidel, *Progressive Pan Americanism: Development and United States Policy toward South America, 1906–1931*, pp. 652–653.

3. The Good Neighbor Policy, 1930–1939

1. Burggraaff, *Venezuelan Armed Forces*, pp. 26–29; Clinton, *Gómez*, p. 257.
2. Eleazar López Contreras, *Gobierno y administración, 1936–1941*, pp. 5–12; John D. Martz, *Acción Democrática: Evolution of a Modern Political Party in Venezuela*, pp. 42–43; Gallegos Ortíz, *La historia política*, pp. 166–167, 185–186.
3. Powell, *Venezuelan Peasant*, pp. 27–28; Summerlin to State Department, August 24, 1934, DSR 831.00/1519; Keeling to Foreign Office, March 3, 1934, FO 371/A7812/7812/47.
4. Lieuwen, *Petroleum*, p. 64.
5. John D. Martz, "Venezuela's Generation of '28': The Genesis of Political Democracy," *Journal of Inter-American Studies* 6 (January 1964): 17–18; McGoodwin to State Department, November 15, 1918, DSR 831.00/865.
6. Martz, "Generation of '28,'" pp. 18–19.
7. Ibid., pp. 19–20; Betancourt, *Venezuela*, pp. 41–44; Burggraaff, *Venezuelan Armed Forces*, pp. 20–24.
8. Martz, *Acción Democrática*, pp. 120–125.
9. Martz, "Generation of '28,'" pp. 23–27.
10. Ibid., pp. 27–30; Powell, *Venezuelan Peasant*, p. 38.
11. Lieuwen, *Petroleum*, pp. 81–82; Burggraaff, *Venezuelan Armed Forces*, pp. 38–40.
12. Martínez, *Torres*, pp. 226–227; Lieuwen, *Petroleum*, pp. 67–68; U.S. Tariff Commission, *Report to the House of Representatives on the Cost of Production of Crude Petroleum*, pp. 2–3.
13. Editorial included in despatch from Minister Meredith Nicholson to State Department, February 24, 1936, DSR 831.6363/844.
14. Nicholson to State Department, February 8, 1936, DSR 831.00/1579, March 28, 1936, DSR 831.00/1611, May 25, 1936, DSR 831.00/1623, June 3, 1936, DSR 831.00B/18, and August 7, 1936, DSR 831.6363/894. Keeling to Foreign Office, April 15, 1936, FO 371/A3738/66/47, July 17, 1936, FO 371/A6451/66/47, and July 24, 1936, FO 371/A6537/66/47. Both Nicholson and Keeling discounted the "red menace" in Venezuela. They believed that Venezuelans wanted better working and living conditions.
15. Nicholson to State Department, August 7, 1936, DSR 831.6363/894.
16. Martz, "Generation of '28,'" pp. 28–30.
17. Analyst quoted in *Fortune* 19 (March 1939): 106; Lieuwen, *Petroleum*,

pp. 81–82; Betancourt, *Venezuela*, pp. 56–58; Burggraaff, *Venezuelan Armed Forces*, pp. 38–40. In the mid-1930s, a bolívar was worth approximately thirty-three cents.

18. Nicholson to State Department, March 16, 1937, DSR 831.5045/50; Lieuwen, *Petroleum*, p. 82.
19. Lieuwen, *Petroleum*, pp. 83–84; *World Petroleum* 10 (December 1939): 52, 68–69; Nicholson to State Department, March 18, 1937, DSR 831.6363/955.
20. Division of Latin American Affairs Accomplishments File, Executive Series, Presidential Papers, Box 49, Hoover Library; Venezuela File, Foreign Affairs—Countries Series, Presidential Papers, Box 994, Hoover Library; Engert to State Department, October 5, 1929, DSR 831.00 General Conditions/11.
21. Lippmann, *World Affairs, 1932*, pp. 49–52; Wilson to State Department, November 25, 1931, DSR 831.00/1498.
22. For Washington's view of the Good Neighbor policy, see April 2, 1938, speech by Laurence Duggan, chief of Division of American Republics, to American Academy of Political and Social Sciences in James W. Gantenbein (ed.), *The Evolution of Our Latin American Policy: A Documentary Record*, pp. 190–193.
23. Keeling to Foreign Office, December 20, 1933, FO 371/A493/225/47.
24. *New York Times*, August 20, 1933, p. 15, and August 22, 1933, p. 16.
25. Nicholson to State Department, June 19, 1936, DSR 831.6363/867.
26. Ibid., May 25, 1936, DSR 831.6363/866, June 19, 1936, DSR 831.6363/867, and July 21, 1936, DSR 831.6363/878.
27. Ibid., June 19, 1936, DSR 831.6363/867, July 21, 1936, DSR 831.6363/878, and September 11, 1936, DSR 831.6363/913.
28. Duggan's comments attached to Nicholson to State Department, June 19, 1936, DSR 831.6363/867; Welles' comments attached to ibid., September 11, 1936, DSR 831.6363/913.
29. Cordell B. Hull, *The Memoirs of Cordell Hull*, I, 350–357; Hull's speech at the special Inter-American Conference on the Maintenance of Peace, Buenos Aires, December 5, 1936, in Gantenbein, *Evolution*, p. 185; Dick Steward, *Trade and Hemisphere: The Good Neighbor Policy and Reciprocal Trade*, pp. 8–12.
30. Hull, *Memoirs*, I, 308; Welles' speech, "The Trade Agreements Program in Our Inter-American Relations," February 2, 1936, in Gantenbein, *Evolution*, p. 171.
31. Quotation from Division of Latin American Affairs Accomplishments File, Vol. II, Executive Series, Presidential Papers, Hoover Library; Hull, *Memoirs*, I, 308; Steward, *Trade and Hemisphere*, pp. 22, 268.
32. *FRUS, 1936*, V, 955–963; memorandum for R. Walton Moore from Henry C. Hawkins, Division of Trade Agreements, November 23, 1936, DSR 611.3131/65½.
33. *World Petroleum* 4 (September 1933): 273–277; Lieuwen, *Petroleum*, pp. 59–60.
34. U.S. Tariff Commission, *Report to the House on Petroleum*, p. 13;

idem, *Foreign Trade of Latin America*, Sec. 10: *Venezuela*, pp. 54–58.

35. Giddens, *Standard Oil Company (Indiana)*, p. 490; Larson et al., *Standard Oil: New Horizons*, pp. 48–49.
36. *FRUS, 1937*, V, 746–795, *1938*, V, 956–969.
37. Ibid., *1938*, V, 971–984; Paul Daniels to State Department, July 15, 1938, DSR 611.3131/226.
38. Daniels to State Department, July 15, 1938, DSR 611.3131/226; for Secretary Hull's response to criticism of the agreement from the domestic oil industry, see *OGJ*, December 7, 1939, p. 25.
39. Lloyd C. Gardner, *Economic Aspects of New Deal Diplomacy*, p. 39; Kenneth J. Grieb, "Negotiating a Reciprocal Trade Agreement with an Underdeveloped Country," *Prologue* 5 (Spring 1973): 22–24.
40. *FRUS, 1938*, V, 960–961.
41. U.S. Department of Commerce, Bureau of Foreign and Domestic Commerce, *Yearbook, 1939*, p. 229; U.S. Department of Commerce, Office of International Trade, *Foreign Commerce Yearbook, 1948*, pp. 412–419; Izard, *Series estadísticas*, pp. 202–203; Steward, *Trade and Hemisphere*, p. 272.
42. Powell, *Venezuelan Peasant*, pp. 22–23, 45–50.
43. Izard, *Series estadísticas*, p. 193; U.S. Department of Commerce, *Commercial and Industrial Development of Venezuela*, p. 14; E. G. Bennion, "Venezuela," in *Economic Problems of Latin America*, ed. Seymour Harris, p. 440; Lieuwen, *Petroleum*, p. 64; Picón-Salas, *Venezuela, independiente*, charts between pp. 416 and 417.
44. Lieuwen, *Petroleum*, p. 73; Chargé d'Affaires John A. MacGregor to Foreign Office, January 27, 1936, FO 371/A1259/66/47.
45. López Contreras, *Gobierno*, pp. 16–30.
46. Salcedo-Bastardo, *Historia fundamental*, pp. 516–518; Lieuwen, *Venezuela*, pp. 73–76.
47. Lieuwen, *Venezuela*, p. 75.
48. U.S. Tariff Commission, *Report to the House on Petroleum*, p. 3; idem, *Report to the Congress on the Costs of Crude Petroleum*, pp. 33–35; *World Petroleum* 14 (June 1943): 31.
49. Lieuwen, *Petroleum*, pp. 78–79; for company admission of tax fraud, see memorandum, Philip Bonsal to Welles, April 8, 1941, DSR 831.6363/1233.
50. Lieuwen, *Petroleum*, p. 79.
51. Lorenzo Meyer, *Mexico and the United States in the Oil Controversy, 1917–1942*, pp. 149–172.
52. Minister Antonio C. Gonzalez to State Department, June 2, 1938, DSR 831.6363/1044; Lieuwen, *Petroleum*, pp. 76–77.
53. Minister of Fomento Manuel Egaña quoted in *World Petroleum*, 11 (September 1940): 42; Gallegos Ortíz, *La historia política*, p. 179.
54. Gonzalez to State Department, June 2, 1938, DSR 831.6363/1044, and September 15, 1938, DSR 831.6363/1062.
55. E. David Cronon, *Josephus Daniels in Mexico*, pp. 151–152, 171, 221.
56. Duggan's memorandum of conversation with Dr. Manuel Egaña, Feb-

ruary 24, 1938, DSR 831.6363/1011; Duggan to Welles, April 6, 1938, DSR 831.6363/1024; Duggan to Hull and Welles, October 23, 1938, DSR 831.6363/1077.

57. Cronon, *Daniels*, pp. 188–198; Bryce Wood, *The Making of the Good Neighbor Policy*, pp. 208–219.
58. Diary entry, March 24, 1938, Vol. 116, 406, Morgenthau Diaries, Franklin D. Roosevelt Library, Hyde Park, New York; Gardner, *Economic Aspects*, p. 118.
59. Duggan's March 17 memorandum is included in Duggan to Hull and Welles, June 26, 1939, DSR 831.6363/1141.
60. Cronon, *Daniels*, pp. 198–202; Wood, *Good Neighbor*, pp. 219–233; Allan S. Everest, *Morgenthau, the New Deal, and Silver*, pp. 86–95; Assistant Secretary of Treasury Harry Dexter White to Morgenthau, March 31, 1939, Folder 14a, Box 6, Harry Dexter White Papers, Firestone Library, Princeton University, Princeton, New Jersey.
61. Scott to State Department, June 15, 1939, DSR 831.6363/1125.
62. Duggan to Hull and Welles, June 26, 1939, DSR 831.6363/1141.
63. Ibid.
64. Hull's memorandum of conversation with Escalante, July 12, 1939, Folder 258, Box 61, Cordell B. Hull Papers.
65. Welles to Scott, July 22, 1939, DSR 831.6363/1125.
66. Frank D. McCann, Jr., *The Brazilian-American Alliance, 1937–1945*, p. 319; entry, March 30, 1942, Vol. 38, Henry L. Stimson Diary, Sterling Memorial Library, Yale University, New Haven, Connecticut.

4. Wartime Policies, 1939–1945

Portions of this chapter first appeared as "Energy for War: United States Oil Diplomacy in Latin America during World War II," in *Proceedings of The Citadel Conference on War and Diplomacy, 1976*, ed. David H. White (Charleston: The Citadel, 1976). I gratefully acknowledge permission to use this material.

1. *FRUS, 1940*, II, 729–730.
2. *Libro amarillo, 1940*, I, I–N, *1941*, I, F–I; Eduardo Plaza A., *La contribución de Venezuela al Panamericanismo, durante el período 1939–1943*, pp. 29–34, 50–53, 213–214; Edith Myretta James Blendon, "Venezuela and the United States, 1928–1948: The Impact of Venezuelan Nationalism" (Ph.D. dissertation), pp. 60–64; Sheldon Liss, *Diplomacy and Dependency: Venezuela, the United States, and the Americas*, pp. 111–112.
3. Alton Frye, *Nazi Germany and the American Hemisphere, 1933–1944*, pp. 21–31; Tattenbach to Foreign Ministry, February 4, 1933, serial 8706/E608478–81 (microfilm), German Foreign Ministry Archives, U.S. National Archives Collection of Captured Foreign Documents (hereafter cited as CFD with appropriate serial and document number). These materials were translated by Genice A. Gladow Rabe. Tattenbach to Foreign Ministry, December 27, 1934, CFD 8705/E608467–74;

Blendon, "Venezuelan Nationalism," p. 68; Liss, *Diplomacy and Dependency*, pp. 106–127.

4. Frye, *Nazi Germany*, p. 101; Ernest Gye to Foreign Office, May 28, 1937, FO 371/A4378/51/47.
5. Legation in Caracas to consulate in Maracaibo, February 9, 1938, DSR 831.00F/3; Scott to State Department, May 27, 1940, DSR 831.00N/19; Ambassador Frank P. Corrigan to State Department, August 16, 1940, DSR 831.00N/34; Paley, "Confidential Memorandum on Internal Conditions and International Relations of Various Latin American Countries," January 7, 1941, South American File, Box 67, President's Secretary's File, Roosevelt Library; Gainer to Foreign Office, March 24, 1941, FO 371/A3969/735/47.
6. Liss, *Diplomacy and Dependency*, pp. 115–120.
7. Paul Daniels to State Department, June 30, 1938, DSR 611.3131/207; Izard, *Series estadísticas*, pp. 202–203.
8. Frye, *Nazi Germany*, pp. 73–74; Frederick Hausermann, "Latin American Oil in War and Peace," *Foreign Affairs* 21 (January 1943): 356; the Reich and Prussian Economic Ministry to Dr. L. S. Rahn of Foreign Ministry, March 9, 1937, CFD 7097/E527121–24; Second Secretary Henry Villard to State Department, February 8, 1937, DSR 831.6363/947; *Libro amarillo, 1937*, p. 107.
9. Poensagen to Foreign Ministry, July 4, 1937, CFD 7096/E527114, and August 6, 1937, CFD 7096/E527116; L. S. Rahn to Poensagen, August 13, 1937, CFD 7096/E527117; *Libro amarillo, 1938*, pp. 357–364; Nicholson to State Department, March 18, 1937, DSR 831.6363/955.
10. *FRUS, 1937*, V, 779–780, *1938*, V, 964–965, and *1943*, VI, 820–854; *Libro amarillo*, p. CIX, 325–333.
11. In his study, *Good Neighbor Diplomacy: United States Policies in Latin America, 1933–1945*, pp. 115–116, Irwin F. Gellman argues that the Roosevelt administration exaggerated the political and economic threat that Nazi Germany posed to the Americas.
12. Stephen G. Rabe, "Inter-American Military Cooperation, 1944–1951," *World Affairs* 137 (Fall 1974): 133–134; Gerald K. Haines, "Under the Eagle's Wing: The Franklin Roosevelt Administration Forges an American Hemisphere," *Diplomatic History* 1 (Fall 1977): 376–378.
13. American Consul Warden McKee Wilson, Genoa, Italy, to State Department, February 7, 1938, DSR 831.34/47; Scott to State Department, September 15, 1939, DSR 831.20/46; Gonzalez to State Department, January 3, 1939, DSR 831/34/53, and despatches DSR 831.34/54–85; Stetson Conn and Byron Fairchild, *The Framework of Hemisphere Defense*, in the official Army history, *United States Army in World War II: The Western Hemisphere*, p. 173.
14. *FRUS, 1940*, V, 177–179, and *1941*, VII, 609–610. The lend-lease treaty signed in March 1942 gave Venezuela $15 million worth of equipment for $6.75 million (*FRUS, 1942*, VI, 739–743). Venezuela received only $4,528,492 worth of matériel because, by the time the United States had surplus equipment to lend, all danger to the Western

Hemisphere had been eliminated. For figures on lend-lease, see U.S. Department of State, *Thirty-Third Report to the Congress on Lend-Lease Operations for the Period Ending December 31, 1951*, p. 34.

15. Frederick C. Adams, *Economic Diplomacy: The Export-Import Bank and American Foreign Policy, 1934–1939*, pp. 156–159, 201–203; Steward, *Trade and Hemisphere*, pp. 253–268; U.S. Department of State, Eleanor Lansing Dulles, *The Export-Import Bank of Washington: The First Ten Years*, pp. 8, 12; Edward O. Guerrant, *Roosevelt's Good Neighbor Policy*, pp. 102, 160.
16. Emilio Collado, executive secretary of Board of Economic Operations of Department of State, to Welles, May 22, 1942, DSR 831.51/311; Corrigan to State Department, November 13, 1939, DSR 831.51A/51; quotations are from Welles' letter to President Roosevelt, July 14, 1939, File 535 (Venezuela), White House Official File, Roosevelt Library; Mr. Hanes to Secretary of Treasury Morgenthau, April 6, 1939, Vol. 176, p. 182, Morgenthau Diaries, Roosevelt Library; Adams, *Export-Import*, p. 205.
17. Fitzhugh Lee, naval attaché in Venezuela, to Rear Admiral John H. Hoover, commander, Caribbean Coastal Frontier, San Juan, Puerto Rico, April 17, 1942, Fitzhugh Lee File, Frank P. Corrigan Papers, Roosevelt Library; *Libro amarillo, 1942*, pp. 32–35; *FRUS, 1945*, IX, 1418–1425.
18. Guerrant, *Good Neighbor*, p. 196; Bennion, "Venezuela," p. 36; Meeting of Senior Members of Joint Military Advisory Board on American Republics, February 3, 1947, Box 1, Record Group 225, Modern Military Records Branch, National Archives.
19. *FRUS, 1942*, VI, 735; comment of Counselor John V. Perowne on despatch, Gainer to Foreign Office, August 11, 1942, FO 371/A7468/503/47.
20. "Radioscript," 1934, Speeches File, Box 26, Corrigan Papers.
21. Corrigan to State Department, October 9, 1939, DSR 831.6363/1155; Assistant Secretary of State Adolf A. Berle to Corrigan, August 26, 1939, DSR 831.6363/1139.
22. Corrigan to State Department, October 9, 1939, DSR 831.6363/1155.
23. Ibid.
24. Memorandum, Livingston Satterthwaite to Duggan, Ellis Briggs, and Collado, September 6, 1939, DSR 831.6363/1148; Duggan to Hull, November 28, 1939, DSR 831.6363/1164. Hull did schedule a meeting with oil company executives for June 1940 (see memorandum, June 7, 1940, DSR 831.6363/1207). There is no record of the meeting. If it took place, Hull, as indicated by the memorandum, would have spoken only in general terms about the need to avoid problems in Venezuela.
25. Hull to Corrigan, December 9, 1939, DSR 611.3131/541A. The Trade Division broke Venezuela's oil allocation into two parts; it assigned 71.9 percent of the quota to oil shipped directly from Venezuela and 20.3 percent to Venezuelan oil refined in the Dutch West Indies (see U.S. Department of State, *Allocation of Tariff Quota on Crude Petroleum and Fuel Oil*).

26. Duggan to Hull, December 9, 1939, DSR 611.3131/541A.
27. Welles to Corrigan, January 2, 1941, DSR 831.6363/1126; Corrigan to State Department, March 18, 1941, DSR 831.6363/1224.
28. Bonsal to Welles, April 8, 1941, DSR 831.6363/1233.
29. Welles to Bonsal, April 8, 1941, DSR 831.6363/1234.
30. Bonsal memorandum, April 19, 1941, DSR 831.6363/1238.
31. Gallegos Ortíz, *La historia política*, pp. 185–186; Salcedo-Bastardo, *Historia fundamental*, pp. 476–478.
32. Burggraaff, *Venezuelan Armed Forces*, pp. 47–49.
33. Ibid., pp. 49–51; Isaías Medina Angarita, *Cuatro años de democracía*, pp. 21–31.
34. Corrigan to State Department, August 12, 1941, DSR 831.6363/1247; Gainer to Foreign Office, June 24, 1941, FO 371/A7435/4224/47.
35. *FRUS, 1942*, VI, 743–744; Gainer to Foreign Office, October 14, 1941, FO 371/A503/503/47; Lieuwen, *Petroleum*, p. 90.
36. Lieuwen, *Petroleum*, pp. 90–91.
37. Ibid., p. 91; Bennion, "Venezuela," p. 447; Enrique A. Baloyra, "Oil Policies and Budgets in Venezuela, 1938–1968," *Latin American Research Review* 9 (Summer 1974): 44.
38. Cronon, *Daniels*, pp. 257–271; Meyer, *Oil Controversy*, pp. 217–225; Wood, *Good Neighbor*, pp. 247–259.
39. Medina quoted in Lieuwen, *Petroleum*, p. 94.
40. Corrigan to State Department, February 25, 1943, DSR 831.6363/1435; Gainer to Foreign Office, August 11, 1942, FO 371/A7468/503/47.
41. *FRUS, 1942*, VI, 746–750; Wood, *Good Neighbor*, p. 272.
42. Corrigan to State Department, February 25, 1943, DSR 831.6363/1435.
43. Ibid., and September 4, 1942, DSR 831.6363/1314; Bonsal's memorandum of conversations between Welles and Thomas Armstrong, Wallace Pratt, and C. H. Lieb of Standard Oil of New Jersey, September 23, 1942, DSR 831.6363/1324.
44. Welles to President Roosevelt, December 30, 1942, File 535 (Venezuela), White House Official File, Roosevelt Library; Cronon, *Daniels*, p. 264; *FRUS, 1942*, VI, 751–752.
45. Gainer to Foreign Office, June 24, 1941, FO 371/A7435/4224/47; Larson et al., *Standard Oil: New Horizons*, p. 480.
46. Nicholson to State Department, September 11, 1936, DSR 831.6363/913.
47. Corrigan to State Department, February 25, 1943, DSR 831.6363/1435.
48. Farish's statement is in letter from F. C. Starling of the Petroleum Department to Perowne of the Foreign Office, September 1, 1942, FO 371/A816/503/47.
49. Halifax to Foreign Office, October 5, 1942, FO 371/A9193/503/47.
50. B. A. Wortley, "The Mexican Oil Dispute, 1938–1946," *Transactions of the Grotius Society* 43 (1959): 26–33; Cronon, *Daniels*, pp. 205–206.
51. Ministry of Fuel and Power to Foreign Office, with attached comments, October 29, 1942, FO 371/A9898/503/47.
52. Gainer to Foreign Office, October 8, 1942, FO 371/A9273/503/47; Star-

ling, Petroleum Department, to Perowne, Foreign Office, November 5, 1942, FO 371/A10253/503/47.

53. Gainer to Foreign Office, June 24, 1941, FO 371/A7435/4224/47, and September 5, 1941, FO 371/A9671/4224/47, and in same folder December 15, 1941, letter of Perowne to M. R. Bridgeman, Petroleum Department.
54. Corrigan to State Department, February 25, 1943, DSR 831.6363/1435; Thornburg to Dean Acheson, chairman of Board of Economic Operations of Department of State, December 27, 1942, DSR 831.6363/1260½.
55. Lieuwen, *Petroleum*, pp. 47–49; Giddens, *Standard Oil (Indiana)*, p. 254.
56. Lieuwen, *Petroleum*, p. 48; Engert to State Department, June 20, 1928, DSR 831.6363/390.
57. For information on Aruba and Curaçao, see U.S. Tariff Commission, *Commercial Policies and Trade Relations of the European Possessions in the Caribbean Area*, pp. 257–282.
58. Corrigan to State Department, February 25, 1943, DSR 831.6363/1435. Aruba and Curaçao had a combined refining capacity of 480,000 barrels a day.
59. Lieuwen, *Petroleum*, pp. 95–97.
60. Tugwell, *Politics of Oil*, pp. 182, 20.
61. Wood, *Good Neighbor*, pp. 167, 334, 278, and particularly chapters 10 and 15. David Green questions Wood's thesis; Green does not, however, analyze the specific episodes upon which Wood bases his interpretation (*The Containment of Latin America*, p. 302 n. 9); see also Adams, *Export-Import*, pp. 224–225 n. 74.
62. Wood, *Good Neighbor*, pp. 346, 353.
63. Baloyra, "Oil Policies," pp. 43–44; Tugwell, *Politics of Oil*, pp. 41–43. Wood admits that he has not analyzed the domestic origins of Medina's policy (*Good Neighbor*, p. 331).
64. U.S. Senate, Subcommittee on Monopoly of Select Committee on Small Business, *International Petroleum Cartel*, 82nd Cong., 2nd sess., p. 165; Larson et al., *Standard Oil: New Horizons*, p. 485; Lieuwen, *Petroleum*, pp. 84, 98–99; Tugwell, *Politics of Oil*, pp. 43–44.
65. Tugwell, *Politics of Oil*, pp. 179–181.
66. Satterthwaite memorandum on conversation with Armstrong, July 3, 1941, DSR 831.6363/1244; Satterthwaite to Assistant Division Chief Walter M. Walmsley and Bonsal, August 12, 1941, DSR 831.6363/1249.
67. Larson et al., *Standard Oil: New Horizons*, pp. 480–482; Mira Wilkins, *The Maturing of Multinational Enterprise: American Business Abroad from 1914 to 1970*, p. 256; Rockefeller quoted in Peter Collier and David Horowitz, *The Rockefellers: An American Dynasty*, pp. 209–212.
68. *OGJ*, March 18, 1943, pp. 24–25; *FRUS, 1944*, VII, 1660–1661.
69. Quotation from Satterthwaite to Walmsley and Bonsal, August 12, 1941, DSR 831.6363/1249; *FRUS, 1943*, VI, 807–809; Lieb, president

of Creole Petroleum, to stockholders, March 11, 1943, DSR 831.6363/1489.

70. For a copy of Pérez Alfonzo's speech, see Rómulo Betancourt, *Venezuela's Oil*, pp. 160–173; for Medina's defense of the law, see his *Cuatro años*, pp. 77–87.
71. Memorandum by First Secretary Rodney Gallop, January 2, 1943, FO 371/A112/94/47; Gainer to Foreign Office, June 28, 1943, FO 371/A6575/1621/47; Petroleum Department to Foreign Office, March 11, 1940, FO 371/A1851/137/47. This folder includes van Hasselt's letter to his superior, Frederick Godber of Shell.
72. Welles to Roosevelt, December 30, 1942, File 535 (Venezuela), White House Official File, Roosevelt Library.
73. Gellman, *Good Neighbor Diplomacy*, p. 56.
74. Oil companies quoted in Special Subcommittee Investigating Petroleum Resources, *American Petroleum*, p. 265; see also survey of industry reaction to the 1943 law by oil analyst Ruth Sheldon in *World Petroleum* 14 (November 1943): 45.
75. Welles to Roosevelt, December 30, 1942, File 535 (Venezuela), White House Official File, Roosevelt Library.

5. The *Trienio*, 1945–1948

1. Betancourt, *Venezuela*, pp. 96–97; Burggraaff, *Venezuelan Armed Forces*, pp. 62–64; Martz, *Acción Democrática*, pp. 56–59.
2. Burggraaff, *Venezuelan Armed Forces*, pp. 65–67; Martz, *Acción Democrática*, pp. 59–60; Salcedo-Bastardo, *Historia fundamental*, pp. 478–480.
3. Burggraaff, *Venezuelan Armed Forces*, pp. 52–60.
4. Ibid., pp. 60–62; Betancourt, *Venezuela*, pp. 93–96; Gallegos Ortíz, *La historia política*, p. 218.
5. Betancourt, *Venezuela*, pp. 99–106.
6. Memorandum by Bainbridge C. Davis for Nelson Rockefeller, assistant secretary of state for Latin American affairs, July 6, 1945, DSR 831.00/7-645; Corrigan to State Department, July 9, 1943, DSR 831.00/1856; *FRUS, 1945*, IX, 1414–1415.
7. *OGJ*, November 3, 1945, p. 63; *FRUS, 1945*, IX, 1401–1417; State Department to Corrigan, October 31, 1945, DSR 831.6363/10-3145; Ambassador George Ogilvie-Forbes to Foreign Office, October 25, 1945, FO 371/AS5588/19/47; Lord Halifax to Foreign Office, October 27, 1945, FO 371/AS5632/19/47, and October 31, 1945, FO 371/AS5828/2386/47; *New York Times*, October 20, 1945, p. 5.
8. Corrigan to Spruille Braden, assistant secretary of state, November 27, 1945, Braden File, Box 1, Corrigan Papers; Corrigan to Dawson, January 2, 1946, Dawson File, Box 3, Corrigan Papers; *FRUS, 1946*, XI, 1298–1302; Corrigan memorandum on Venezuelan political situation, December 13, 1946, DSR 831.00/11-1346; *FRUS, 1947*, VIII, 131–144, 1054–1064; Corrigan to Braden, May 13, 1947, Braden File, Box 1, Corrigan Papers; Norman Armour, assistant secretary of state, memoran-

dum of conversation with Corrigan and Proudfit, July 11, 1947, DSR 831.6363/7-1147.

9. Rabe, "Inter-American Military Cooperation," pp. 132–149.
10. *FRUS, 1946*, XI, 1306–1322; Joseph Flack, Office of North and West Coast Affairs, memorandum, January 28, 1946, DSR 831.248/1-2846; Corrigan to State Department, May 2, 1946, DSR 831.00/5-246; Dawson to State Department, September 24, 1946, DSR 831.24/9-2446.
11. Martz, *Acción Democrática*, pp. 250–252; "Policy and Information Statement for Venezuela," July 1, 1946, Papers of James F. Byrnes, Robert M. Cooper Library, Clemson University, Clemson, South Carolina; Dawson to State Department, October 22, 1946, DSR 831.00/10-2246.
12. Randall Bennett Woods, *The Roosevelt Foreign Policy Establishment and the "Good Neighbor": The United States and Argentina, 1941–1945*, pp. 79–130.
13. Beatrice Bishop Berle and Travis Beale Jacobs (eds.), *Navigating the Rapids, 1918–1971: From the Papers of Adolf A. Berle*, pp. 470–475; Spruille Braden, *Diplomats and Demagogues: The Memoirs of Spruille Braden*, pp. 316–338, 356–370; U.S. Department of State, *Consultation among the American Republics with Respect to the Argentine Situation.*
14. *FRUS, 1946*, XI, 220, and *1945*, IX, 185–196; Braden to Samuel Inman, December 6, 1945, Box 15, Papers of Samuel Guy Inman, Library of Congress; *Libro amarillo, 1947*, pp. xii–xiii.
15. Green, *Containment of Latin America*, pp. 276–283.
16. Betancourt quoted in Corrigan to State Department, March 17, 1947, DSR 831.00/3-1747; Corrigan to State Department, December 2, 1946, DSR 831.6363/12-2446; Robert J. Alexander, *The Communist Party in Venezuela*, pp. 24–25; Betancourt, *Venezuela*, pp. 148–150.
17. *OGJ*, May 20, 1948, p. 125.
18. Betancourt quoted in Tugwell, *Politics of Oil*, p. 45; Baloyra, "Oil Policies," pp. 45–47.
19. Juan Pablo Pérez Alfonzo, *El pentágono petrolero*, pp. 102–106; *FRUS, 1946*, XI, 1330–1333, 1344–1346; Dawson to State Department, January 10, 1946, DSR 831.6363/1-1046, and January 25, 1946, DSR 831.6363/1-2546; Secretary Thomas Maleady to State Department, April 27, 1946, DSR 831.6363/4-2746, and November 20, 1946, DSR 831.6363/11-2047.
20. *FRUS, 1946*, XI, 1333–1343; Dawson to State Department, January 10, 1946, DSR 831.6363/1-1046.
21. *FRUS, 1946*, XI, 1344–1346; Secretary Thomas Mann to State Department, December 14, 1948, DSR 831.6363/12-1448.
22. Maleady to State Department, April 27, 1946, DSR 831.6363/10-2345, *FRUS, 1946*, XI, 1344–1346; Wilkins, *Maturing of Multinational Enterprise*, p. 317.
23. R. Townsend, Petroleum Department, memorandum of conversation with Proudfit, June 24, 1946, DSR 831.6363/6-2446; Milton K. Wells, Office of North and West Coast Affairs, memorandum of meeting with

Standard Oil Executives, June 24, 1946, DSR 831.24/6-2446; Corrigan to State Department, December 2, 1946, DSR 831.6363/12-246; Maleady to State Department, November 20, 1947, DSR 831.6363/11-2047.

24. Collier and Horowitz, *The Rockefellers*, pp. 261–265; Oral History of Francis Jamieson, pp. 175–182, Columbia University Oral History Project, New York, New York. Jamieson served on the board of directors of the corporation.
25. Tugwell, *Politics of Oil*, p. 179; *FRUS, 1948*, IX, 763–764; "Creole Petroleum: Business Embassy," *Fortune* 39 (February 1949): 91–93, 178, 180; see also Appendix, Table C.
26. *OGJ*, July 13, 1946, p. 68, and January 29, 1948, p. 143; Betancourt, *Venezuela*, pp. 124–151; Lieuwen, *Petroleum*, pp. 106–108; Maleady to State Department, October 3, 1947, DSR 831.6363/10-347; *FRUS, 1948*, IX, 756–759.
27. *FRUS, 1946*, XI, 1331. The quotation is from Secretary Byrnes' despatch to Chargé Dawson.
28. Secretary of war to secretary of state, June 4, 1946, OPD 92 TS, Modern Military Records Branch, U.S. National Archives; Military Intelligence Division, "Estimate of World Situation and Its Military Implications for the United States," August 16, 1946, OPD 350-05 TS, Modern Military Records, Dawson to State Department, January 19, 1946, DSR 831.00/1-1946; Wells memorandum, May 27, 1946, DSR 831.6363/5-2746.
29. *FRUS, 1946*, XI, 1346–1348; Wells to Braden and Ellis Briggs, July 27, 1946, DSR 831.00B/7-2346; Mr. Hussey to Wells, October 17, 1946, DSR 831.00B/10-1746; Ambassador Walter Donnelly to State Department, December 18, 1947, DSR 831.00 Betancourt, Rómulo/12-1847; *FRUS, 1948*, IX, 194–201, 760–761, 766–770.
30. U.S. Senate, Special Subcommittee Investigating Petroleum Resources, *American Petroleum*, pp. 1–11; U.S. House, Special Subcommittee on Armed Services, *A Report of Investigation of Petroleum in Relation to National Defense*, 80th Cong., 2nd sess., pp. 6053–6066; U.S. Department of State, "Current and Prospective World-Wide Petroleum Situation," *Department of State Bulletin*, March 28, 1948, pp. 426–427 (hereafter cited as *DSB* with appropriate date and page); Piggot, Petroleum Department, to Loftus, Petroleum Department, January 21, 1947, DSR 831.6363/1-2147; *FRUS, 1948*, IX, 243–257; diary entries, January 6, 1948, and January 16, 1948, vol. 9, February 22, 1949, vol. 14, Papers of James V. Forrestal, Firestone Library, Princeton University, Princeton, New Jersey.
31. Wells memorandum of conversation with War and Navy departments, July 17, 1947, DSR 831.00/7-1747; Armour memorandum of conversation with Corrigan and Proudfit, July 11, 1947, DSR 831.6363/7-1147.
32. Martz, *Acción Democrática*, pp. 225–252; Betancourt, *Venezuela*, pp. 165–167.
33. Powell, *Venezuelan Peasant*, pp. 44–50; Picón-Salas, *Venezuela independiente*, pp. 416–417; Clifton R. Wharton, "C.B.R. in Venezuela," *Inter-American Economic Affairs* 4 (Winter 1950): 3; Agricultural At-

taché James H. Kempton to State Department, July 13, 1949, DSR 831.504/7-1349, and July 18, 1949, DSR 831.6363/7-1849.

34. Martz, *Acción Democrática*, pp. 83–84; Burggraaff, *Venezuelan Armed Forces*, pp. 80–81.
35. Dean Acheson to Corrigan, October 7, 1944, DSR 831.24/8-2344; *FRUS, 1943*, V, 291–307.
36. *FRUS, 1944*, VII, 1644–1660; Philip Bonsal memorandum, February 5, 1944, DSR 831.001 Medina (Angarita) Isaías/143; Corrigan to State Department, May 16, 1944, DSR 831.021/88, and September 28, 1945, DSR 631.003/9-2845.
37. Betancourt, *Venezuela*, pp. 192–196; Blendon, "Venezuela and the United States," pp. 148–150.
38. *FRUS, 1947*, VIII, 554.
39. Maleady to State Department, June 13, 1947, DSR 800.85/4-2446, and September 12, 1947, DSR 831.85/9-1247; Donnelly to State Department, December 16, 1947, DSR 831.001; Betancourt, Rómulo/12-1647; Betancourt, *Venezuela*, p. 195.
40. Betancourt quoted in despatch, Donnelly to State Department, December 16, 1947, DSR 831.001 Betancourt, Rómulo/12-1647; *FRUS, 1947*, VIII, 565, and *1950*, II, 1019–1024; R. R. Adams, president of Grace Lines, Inc., to Dean Acheson, October 25, 1950, "G" Folder, Box 2, Papers of Edward R. Miller, Jr., Harry S Truman Library, Independence, Missouri; *Libro amarillo, 1954*, pp. 62–64.
41. U.S. Department of Commerce, Office of International Trade, *Foreign Commerce Yearbook, 1950*, pp. 681–688; U.S. Department of Commerce, Bureau of Foreign Commerce, World Trade Information Service, *Foreign Trade of Venezuela, 1956–1957*; see also Appendix, Table A.
42. Green, *Containment of Latin America*, pp. 169–183; speeches by Assistant Secretary of State Norman Armour, Secretary of State George Marshall, Director of Office of American Republic Affairs Paul Daniels in Gantenbein (ed.), *Evolution*, pp. 271–292; Spruille Braden, "Latin American Industrialization and Foreign Trade," in *The Industrialization of Latin America*, ed. Lloyd J. Hughlett, pp. 487–493.
43. Stephen G. Rabe, "The Elusive Conference: United States Economic Relations with Latin America, 1945–1952," *Diplomatic History* 2 (Summer 1978): 279–294.
44. Zettler and Cutler, "Direct Investments in Foreign Countries," pp. 8–9; see also Appendix, Table B.
45. Charles F. Knox, Office of North and West Coast Affairs, to Corrigan, October 8, 1946, Charles F. Knox, Jr. File, Box 6, Corrigan Papers; Marshall to embassy, July 2, 1948, DSR 831.6363/7-248.
46. Acheson to embassy, April 7, 1947, DSR 831.6363/4-747; embassy to State Department, April 14, 1947, DSR 831.6363/4-1447; Dawson to Corrigan, April 24, 1947, Dawson File, Box 3, Corrigan Papers.
47. *FRUS, 1948*, IX, 761–769; Betancourt, *Venezuela*, pp. 203–207; John C. Rayburn, "Development of Venezuela's Iron-Ore Deposits," *Inter-American Economic Affairs* 6 (Winter 1952): 52–70.

48. Memorandum on Venezuela by Robert Post, Office of North and West Coast Affairs, to White House, June 26, 1948, DSR 831.00/7-1448.
49. Pérez Alfonzo, *El pentágono*, pp. 7, 73–74, 83–88; Tugwell, *Politics of Oil*, pp. 33–37; Vallenilla, *Oil*, pp. 111–119.
50. Salcedo-Bastardo, *Historia fundamental*, p. 481; Burggraaff, *Venezuelan Armed Forces*, pp. 79–89; Levine, *Conflict and Political Change*, pp. 36–41; Martz, *Acción Democrática*, pp. 81–89; Powell, *Venezuelan Peasant*, pp. 79–86.
51. Quoted in Martz, *Acción Democrática*, p. 306.
52. Quotation from ibid., p. 81; Burggraaff, *Venezuelan Armed Forces*, pp. 89–111; Donnelly to State Department, November 17, 1948, DSR 831.00/11-1748.
53. Memorandum on Venezuela by Post to White House, June 26, 1948, DSR 831.00/7-1448; memorandum by Post to Office of American Republic Affairs, June 4, 1948, DSR 831.00/6-448; Under Secretary of State Robert Lovett to Charles Sawyer, secretary of commerce, September 24, 1948, DSR 831.6511/9-2448.
54. *FRUS, 1948*, IX, 134–136, and *1949*, II, 795–797.
55. *FRUS, 1948*, IX, 141–152; U.S. Department of State, "U.S. Concerned at Overthrow of Governments of Certain American Republics," *DSB*, January 2, 1949, p. 30.
56. Baloyra, "Oil Policies," pp. 45–49; Lieuwen, *Petroleum*, pp. 103–110.
57. *FRUS, 1948*, IX, 136–139, 144–145, 151. Gallegos later retracted his charge.
58. Creole Petroleum Corporation, "Consideration on the Effect of the Oil Industry on Venezuela's Economy," 1947, Correspondence File, Numeric File I, 6000-Mining, Box 1, Papers of Merwin L. Bohan, Truman Library; *The Lamp* 30 (January 1948): 2–7; "Creole Petroleum," *Fortune*, p. 179; Secretary Thomas Mann to State Department, December 14, 1948, DSR 831.6363/12-1448.

6. Cold War Policies, 1949–1958

1. Burggraaff, *Venezuelan Armed Forces*, pp. 113–121; Glen L. Kolb, *Democracy and Dictatorship in Venezuela, 1945–1958*, pp. 73–75.
2. Kolb, *Democracy and Dictatorship*, pp. 76–92; Burggraaff, *Venezuelan Armed Forces*, pp. 123–125; Walter Donnelly to State Department, June 9, 1949, DSR 831.001 Delgado Chalbaud/6-949.
3. Kolb, *Democracy and Dictatorship*, pp. 93–108; Burggraaff, *Venezuelan Armed Forces*, pp. 123–125; Salcedo-Bastardo, *Historia fundamental*, pp. 482–483.
4. Martz, *Acción Democrática*, p. 140; Kolb, *Democracy and Dictatorship*, pp. 109–126; Burggraaff, *Venezuelan Armed Forces*, pp. 125–129; Laureano Vallenilla Lanz, *Escrito de memoria*, pp. 167–180.
5. Donnelly to State Department, "Memorandum of conversation with Delgado Chalbaud," May 4, 1949, DSR 831.00/5-449; memorandum by W. L. Krieg, Division of North and West Coast Affairs, July 5, 1949,

DSR 831.00/7-549; Secretary Thomas Mann to Donnelly, June 21, 1949, DSR 831.00/6-2149; *FRUS, 1950,* II, 592.

6. "Department of State Policy Statement," June 30, 1950, *FRUS, 1950,* II, 1039.
7. Donnelly to State Department, November 17, 1948, DSR 831.00/11-1748; Mann to State Department, June 9, 1949, DSR 831.00B/6-949.
8. *FRUS, 1949,* II, 797–801; memorandum by Krieg of conversation with Venezuelan Chargé d'Affaires Dr. Santigo, March 25, 1949, DSR 831.00/3-2549.
9. *FRUS, 1950,* II, 1039; Embassy of Venezuela, Washington, D.C., *Venezuela Up-To-Date* 2 (April 1951): 11 (hereafter cited as *Venezuela* with appropriate volume, date, and page); *Libro amarillo, 1948–1952,* pp. 172–176.
10. *FRUS, 1950,* II, 1025–1026.
11. *FRUS, 1948,* IX, 769–770; Under Secretary Robert Lovett to embassy in Caracas, October 21, 1948, DSR 831.6363/10-2148; Secretary of State Marshall to embassy, December 28, 1948, DSR 831.6363/12-2848; *FRUS, 1949,* II, 795–797; Secretary of State Acheson to embassy, June 16, 1949, DSR 831.6363/6-1649. Philip Agee, a former agent of the CIA, alleged that the agency was still screening employees for left-wing connections in 1960. Agee stated that such services were typically provided for large U.S. corporations in Latin America. Exxon, the parent company of Creole Petroleum of Venezuela, denied the allegation. See *New York Times,* January 13, 1975, p. 8.
12. *Libro amarillo, 1948–1952,* pp. xxviii–xxix, 50–51; John M. Thompson, third secretary of embassy, to State Department, August 18, 1948, DSR 831.00B/8-1848; Donnelly to State Department, April 9, 1949, DSR 831.00B/4-949; *FRUS, 1950,* II, 1029–1030; *Venezuela* 3 (July–August 1952): 19.
13. Under Secretary of State James Webb to James Loy, executive secretary of National Security Council, November 2, 1950, *FRUS, 1950,* II, 670, 624 n. 3.
14. Miller quoted in U.S. Congress, Senate Foreign Relations Committee and Armed Services Committee, *Hearings on the Mutual Security Act of 1951,* 82nd Cong., 1st sess., p. 404; *DSB,* March 30, 1953, pp. 463–467; U.S. Agency for International Development, *U.S. Overseas Loans and Grants and Assistance from International Organizations: Obligations and Loan Authorizations, July 1, 1945–June 30, 1973,* p. 61.
15. Donnelly to State Department, July 30, 1948, DSR 831.6363/7-3048; Baloyra, "Oil Policies," pp. 47–49; Vallenilla, *Oil,* pp. 69–70; *OGJ,* June 2, 1949, p. 111; *Venezuela* 1 (May 1950): 22–23.
16. Joseph E. Pogue, *Oil in Venezuela.*
17. *OGJ,* April 7, 1949, p. 53, June 2, 1949, p. 111, and February 2, 1950, p. 34; Mann to State Department, July 13, 1949, DSR 831.6363/7-1349.
18. Memorandum by H. B. Clark of conversation with Terry Duce, vice-president of ARAMCO, October 15, 1949, DSR 831.6363/10-549; Dorsy,

embassy in Bagdad, to State Department, November 7, 1949, DSR 831.6363/11-749; Richards, embassy in Tehran to State Department, December 1, 1949, DSR 831.6363/2-149.

19. *OGJ*, January 21, 1952, p. 65; Donnelly to State Department, September 29, 1949, DSR 831.6363/9-2949; *FRUS, 1950*, II, 1042–1046; *OGJ*, December 8, 1952, p. 83.
20. *OGJ*, November 2, 1950, pp. 44–45.
21. *FRUS, 1950*, II, 1039; Wilkins, *Maturing of Multinational Enterprise*, p. 290.
22. *FRUS, 1950*, II, 1026–1027; U.S. Department of Commerce, Office of International Trade, *Investment in Venezuela: Conditions and Outlooks for United States Investors*, pp. 14–15; W. Tapley Bennett, deputy director, Office of South American Affairs, "Economic Structure of Pan Americanism," *DSB*, August 11, 1952, pp. 208–209; Krieg to Miller, October 7, 1949, DSR 831.6363/9-3049; memorandum by Robert Eakens, chief of Petroleum Section, on meeting with Minister Egaña, October 13, 1949, DSR 831.6363/10-1349.
23. U.S. Department of State, *Analysis of the Reciprocal Concessions and General Provisions of the Supplementary Trade Agreement between the United States and Venezuela, Signed at Caracas, Venezuela, August 18, 1952*; President Truman to Congressman Omar Burleson of Texas, November 5, 1952, Official File 61, Truman Library; *Libro amarillo, 1948–1952*, pp. cxiv–cxxvi; Cabot quoted in *Venezuela* 4 (April 1953): 7.
24. U.S. Department of State, Office of Public Affairs, *Venezuela: Oil Transforms a Nation*.
25. Burggraaff, *Venezuelan Armed Forces*, pp. 114–115; Donnelly to State Department, June 4, 1948, DSR 831.00/6-448, and June 11, 1948, DSR 831.00/6-1148.
26. Marcos Pérez Jiménez, *Diez años de desarrollo*, pp. 30–58; Burggraaff, *Venezuelan Armed Forces*, pp. 103–132; Kolb, *Democracy and Dictatorship*, pp. 148–168; Tad Szulc, *Twilight of the Tyrants*, p. 249.
27. Burggraaff, *Venezuelan Armed Forces*, pp. 132–135; Pérez Jiménez, *Diez años*, pp. 59–104.
28. Philip B. Taylor, Jr., *The Venezuelan Golpe de Estado of 1958: The Fall of Marcos Pérez Jiménez*, pp. 36–39.
29. Kolb, *Democracy and Dictatorship*, pp. 148–151; *Venezuela* 7 (June 1957): 5; Powell, *Venezuelan Peasant*, pp. 90–97.
30. Szulc, *Twilight*, p. 10; Salcedo-Bastardo, *Historia fundamental*, p. 484.
31. Oral History of John C. Dreier, p. 10, John Foster Dulles Oral History Collection, Firestone Library, Princeton University; diary entry, February 8, 1955, Papers of Adolf A. Berle, Roosevelt Library; Oral History of José Figueres, president of Costa Rica, pp. 32–35, Truman Library; Louis A. Perez, "International Dimensions of Inter-American Relations," *Inter-American Economic Affairs* 27 (Summer 1973): 50–53.
32. Alexander, *Communist Party in Venezuela*, pp. 24–30.

33. Estrada quoted in *Venezuela* 5 (December 1954): 7; U.S. General Services Administration, *Public Papers of the President: Dwight D. Eisenhower, 1955,* 228–229.
34. Memorandum for the president from John Foster Dulles, September 29, 1956, White House Official File, Dwight D. Eisenhower Library, Abilene, Kansas; Council on Foreign Relations, Richard P. Stebbins, *The United States in World Affairs, 1954,* pp. 369–377.
35. Agency for International Development, *Overseas Loans and Grants,* p. 61; *Venezuela* 7 (July–August 1956): 5. After Pérez Jiménez' overthrow, the two countries negotiated a different military aid agreement.
36. The award's announcement is reprinted in Kolb, *Democracy and Dictatorship,* pp. 142–143; for Ambassador Warren's attitudes, see Szulc, *Twilight,* p. 11; for award from navy, see *Venezuela* 8 (December 1957): 3.
37. Kolb, *Democracy and Dictatorship,* pp. 142–143.
38. For views on foreign economic policy, see respective speeches by Assistant Secretary of State Henry Holland and Humphrey in *DSB* 31 (November 1954): 684–690, and 31 (December 1954): 863–869; Oral History of John Cabot, political advisor for Latin American Affairs, pp. 2–13, Dulles Oral History Collection.
39. "Reflections on United States Economic Cooperation with Latin America," March 1968, Reference File: Studies and Reports, Box 14, Merwin Bohan Papers, Truman Library; Bohan to Holland, October 26, 1954, White House Official File 79-A-2 (Inter-American Economic and Social Council), Eisenhower Library.
40. For figures on investment, see U.S. Senate, Committee on Foreign Relations, Subcommittee on Multinational Corporations, *Multinational Corporations in Brazil and Mexico: Structural Sources of Economic and Noneconomic Power,* 94th Cong., 1st sess., pp. 34–37; for data on trade, see Banco Central de Venezuela, *La economía venezolana en los últimos trienta años,* pp. 129, 131; see also Appendix, Tables A and B.
41. Tugwell, *Politics of Oil,* pp. 179; Wayne C. Taylor and John Lindeman, *The Creole Petroleum Corporation in Venezuela,* pp. 2–3.
42. Quoted in Betancourt, *Venezuela,* p. 356; for a history of iron ore development, see *Venezuela* 9 (December 1959): 7–10.
43. Tugwell, *Politics of Oil,* pp. 47–49; Baloyra, "Oil Policies," pp. 47–49.
44. Creole Petroleum Corporation, *Data on Petroleum and Economy of Venezuela, 1974,* p. 12; *Venezuela* 7 (May–June 1956): 17, and 8 (March 1958): 8.
45. Taylor and Lindeman, *Creole Petroleum,* pp. 33–38, 57–58, 71; *Venezuela* 6 (January 1956): 11–12, 7 (December 1956): 8, and 7 (January 1957): 16.
46. Szulc, *Twilight,* pp. 252–253; for speeches by U.S. business community praising Pérez Jiménez, see respective remarks by First National City Bank, U.S. Steel, Creole Petroleum, and the U.S. Chamber of Commerce of Venezuela reprinted in *Venezuela* 8 (November 1957): 3–4, 7.

47. Humphrey's remarks in *Venezuela* 5 (January 1955): 7; Holland's statement in U.S. Senate, Foreign Relations Committee, *Hearings on the Mutual Security Act of 1956*, 84th Cong., 2nd sess., p. 308; Dulles' testimony in U.S. Senate, Committee on Finance, *Hearings on H.R. 1 (Trade Agreements Extension Act)*, Part 4, 84th Cong., 1st sess., p. 2049.
48. Council on Foreign Relations, Richard P. Stebbins, *The United States in World Affairs, 1958*, p. 360; Celso Furtado, *Economic Development of Latin America: A Survey from Colonial Times to the Cuban Revolution*, pp. 46–48; Martz, *Acción Democrática*, p. 91; Samuel L. Baily, *The United States and the Development of South America, 1945–1975*, p. 6.
49. Banco Central, *La economía venezolana*, pp. 54, 56, 63; Powell, *Venezuelan Peasant*, pp. 87–94; U.S. Department of Agriculture, *U.S. Farm Products Find Market and Competition in Venezuela*, pp. 16, 19; Levine, *Conflict and Political Change*, p. 17; Burggraaff, *Venezuelan Armed Forces*, pp. 141–142.
50. Betancourt, *Venezuela*, pp. 320–359; Creole Petroleum, *Data, 1974*, p. 28; Banco Central, *La economía venezolana*, p. 93.
51. Taylor, *Venezuelan Golpe de Estado*, p. 50; Powell, *Venezuelan Peasant*, pp. 98–99.
52. Vallenilla Lanz, *Escrito de memoria*, pp. 225–230; Burggraaff, *Venezuelan Armed Forces*, pp. 147–149; Martz, *Acción Democrática*, pp. 93–94.
53. Burggraaff, *Venezuelan Armed Forces*, pp. 143–147, 149–154, 212; Taylor, *Venezuelan Golpe de Estado*, pp. 37–38; Martz, *Acción Democrática*, pp. 94–95.
54. Burggraaff, *Venezuelan Armed Forces*, pp. 154–168.
55. Ibid., pp. 169–189.
56. Pérez Jiménez' speech in *Venezuela* 7 (August–September 1956): 3–4; Taylor, *Venezuelan Golpe de Estado*, p. 2; Domingo Alberto Rangel, *La revolución de las fantasías*, pp. 42–43, 198–199.
57. *DSB*, March 31, 1958, p. 520.
58. *Venezuela* 8 (March 1958): 4, 15; *OGJ*, February 3, 1958, p. 52; Berle and Jacobs, *Navigating the Rapids*, p. 684.
59. Richard M. Nixon, *Six Crises*, pp. 195–247; Rangel, *La revolución*, pp. 202–209.
60. Waugh to family, May 26, 1958, Correspondence File, 1954–1958, Box 1, Papers of Samuel Waugh, Eisenhower Library.
61. Baily, *Development of South America*, pp. 75–79; Stebbins, *The United States in World Affairs, 1958*, pp. 351–361.
62. Oral History of C. Douglas Dillon, pp. 35–41, Dulles Oral History Collection; Oral History of Roy Richard Rubottom, assistant secretary of state, pp. 30–59, Dulles Collection; Oral History of Milton Eisenhower, pp. 7–8, 95–101, Columbia University Collection; Milton Eisenhower, *The Wine Is Bitter*, pp. 187–254.
63. U.S. Senate, *Mutual Security Act of 1956*, pp. 284–285; U.S. Senate,

Committee on Foreign Relations, *Hearings on the Mutual Security Act of 1959*, 86th Cong., 1st sess., pp. 543–551; U.S. Senate, Committee on Foreign Relations, Senator Wayne Morse, *Report on Study Trip to South America*, 86th Cong., 1st sess., pp. 33–34.

64. Guggenheim to President Eisenhower, May 26, 1958, and Fleming to Eisenhower, November 19, 1958, White House Official File 116-J-Latin America, Folder 4, Eisenhower Library; see also papers of National Advisory Committee on Inter-American Affairs, White House Official File 116-J-12, Eisenhower Library.
65. Nixon, *Six Crises*, p. 205.
66. Newspaper clipping from *Caracas Journal*, August 18, 1958, Box 893, White House Official File 227 (Venezuela), Eisenhower Library; Judith Ewell, "The Extradition of Marcos Pérez Jiménez, 1959–1963: Practical Precedent for Enforcement of Administrative Honesty," *Journal of Latin American Studies* 9 (November 1977): 291–313.
67. Baily, *Development of South America*, pp. 77–79; Jerome Levinson and Juan de Onís, *The Alliance That Lost Its Way: A Critical Report on the Alliance for Progress*, pp. 42–49.
68. *DSB*, June 10, 1958, p. 1107; Nixon quoted in Stebbins, *The United States in World Affairs, 1958*, p. 354.
69. Department quoted in Berle Diary, entry of July 9, 1958, Roosevelt Library; Nixon, *Six Crises*, p. 238.
70. Diary entries, October 16, 1958, November 24, 1958, December 6, 1958, Berle Papers, Roosevelt Library.
71. Memorandum to President Eisenhower from Herter, December 17, 1958, White House Official File 227 (Venezuela), Box 893, Eisenhower Library; General Services Administration, *Public Papers of the President: Eisenhower, 1958*, p. 866.
72. Diary entry, October 16, 1958, Berle Papers, Roosevelt Library; Berle and Jacobs, *Navigating the Rapids*, p. 666.
73. Herter to Eisenhower, December 17, 1958, White House Official File 227 (Venezuela), Box 893, Eisenhower Library; *OGJ*, December 15, 1958, p. 84.
74. Dulles quoted in Council on Foreign Relations, Richard P. Stebbins, *The United States in World Affairs, 1957*, pp. 259–260.
75. Levinson and de Onís, *Alliance*, pp. 6, 42–49; Sherman Adams, assistant to president, to Senator Homer Capeheart, March 25, 1958, White House General File 122-G-Latin America, Eisenhower Library; memorandum of conversation between Eisenhower and Senator Bourke Hickenlooper, September 13, 1960, White House Official File 116-J-Latin America, Folder 5, Eisenhower Library.
76. Eisehower quoted in *Venezuela* 10 (January–February 1961): 4–6.

7. The Alliance for Progress, 1959–1968

1. *Venezuela* 9 (January 1959): 3–4. Betancourt's book has been translated: *Venezuela: Oil and Politics*.
2. Martz, *Acción Democrática*, p. 110; John R. Dinkelspiel, "Technology

and Tradition: Regional and Urban Development in the Guayana" in *Venezuela, 1969: Analysis of Progress*, ed. Philip B. Taylor, Jr., pp. 122–145; Powell, *Venezuelan Peasant*, pp. 106–114.

3. Levine, *Conflict and Political Change*, pp. 44, 109–110.
4. Robert J. Alexander, *Latin American Political Parties*, pp. 338–354; Franklin Tugwell, "The Christian Democrats of Venezuela," *Journal of Inter-American Studies* 7 (April 1965): 254, 258–262.
5. Edwin Lieuwen, *Generals vs. Presidents: Neo-Militarism in Latin America*, pp. 87–89; Burggraaff, *Venezuelan Armed Forces*, pp. 201–203.
6. Martz, *Acción Democrática*, pp. 177–183; Alexander, *Communist Party in Venezuela*, pp. 70–72; Powell, *Venezuelan Peasant*, pp. 112–114.
7. Spokesperson of MIR quoted in Richard Gott, *Guerilla Movements in Latin America*, p. 136.
8. Council on Foreign Relations, Richard P. Stebbins, *The United States in World Affairs, 1960*, pp. 310–315, 328–329, and *The United States in World Affairs, 1961*, pp. 295–296; Liss, *Diplomacy and Dependency*, pp. 231–239.
9. Levinson and de Onís, *Alliance*, pp. 5–16, 34–36, 64–73; Abraham F. Lowenthal, "United States Policy toward Latin America: 'Liberal,' 'Radical,' and 'Bureaucratic' Perspectives," *Latin American Research Review* 8 (Spring 1974): 3–25; Dillon quoted in Stebbins, *World Affairs, 1961*, p. 326.
10. Levinson and de Onís, *Alliance*, pp. 49–56; Arthur M. Schlesinger, Jr., *A Thousand Days: John F. Kennedy in the White House*, pp. 189–190.
11. Memorandum [draft] for President-elect Kennedy from Berle, December 13, 1960, and Berle to Richard Goodwin, office of Senator Kennedy, January 9, 1961, Latin American Task Force File, Box 94, Berle Papers; diary entries, November 14, 1960, December 8, 1960, and February 27, 1961, Berle Papers; Arthur Schlesinger to Kennedy, November 11, 1960, Staff Memoranda, Schlesinger File, November 1960–June 1961, Box 65, President's Office Files, John F. Kennedy Library, Boston, Massachusetts.
12. Schlesinger, *Thousand Days*, p. 185; diary entries, October 25, 1959, and February 23, 1961, Berle Papers; Berle to Goodwin, January 9, 1961, Latin American Task Force File, Box 94, Berle Papers; Betancourt, *Venezuela*, pp. 148–150; speech by Betancourt in *Venezuela* 10 (April–May 1960): 3.
13. Schlesinger, *Thousand Days*, p. 202; Levinson and de Onís, *Alliance*, pp. 56–58; speech by Mayobre to Pan American Society, New York, February 21, 1961, Latin American Folder (1/61–2/61), Box 215, National Security Council Files, Kennedy Library (hereafter cited as NSF with appropriate folder and box); Berle to Kennedy, "Report of the Task Force on Latin America," July 7, 1961, Box 220, Berle Diary.
14. Schlesinger, *Thousand Days*, p. 766; memorandum for Kennedy from Schlesinger, March 10, 1961, NSF: Latin American Folder (3/8/61–

3/14/61), Box 215; Rostow to Kennedy, March 2, 1961, NSF: Latin American Folder (3/1/61–3/7/61), Box 215. A key exposition in the "middle-class revolution" thesis is John Johnson's *Political Change in Latin America: The Growth of the Middle Sectors.* Robert Alexander, who served on the task force for Latin America, subscribed to Johnson's thesis; see Alexander's *The Venezuelan Democratic Revolution: A Profile of the Regime of Rómulo Betancourt.*

15. Betancourt, *Venezuela*, p. 389; Rómulo Betancourt, *Venezuela and the U.S. Alliance for Progress*, p. 2.
16. General Services Administration, *Public Papers of the President: John F. Kennedy, 1963*, pp. 184–186.
17. *DSB*, May 22, 1961, p. 764; Secretary of State Rusk to Teodoro Moscoso, May 29, 1961, NSF: CO: Venezuela, General (1/61–6/61), Box 192; Stewart to State Department, July 2, 1962, NSF: CO: Venezuela, General (2/1/62–8/19/62), Box 197.
18. Memorandum, "Background Papers on Dr. Rafael Caldera," for Ralph Dungan, White House, from William H. Brubeck, executive secretary of National Security Council, June 27, 1962, NSF: CO: Venezuela, General (6/5/62–6/30/62), Box 192; memorandum of meeting between Kennedy and General Antonio Briceño Linares, August 30, 1962, NSF: CO: Venezuela, General (9/1/62–9/20/62), Box 192; Stewart to State Department, June 4, 1962, NSF: CO: Venezuela, General (6/1/62–6/4/62), Box 192.
19. Lyndon Johnson to Raúl Leoni, July 9, 1966, CO 311 (Venezuela), Box 79, General Files, White House Central Files, Lyndon B. Johnson Library, Austin, Texas.
20. Betancourt to Kennedy, October 25, 1961, Country File (Venezuela), Box 128, President's Office File, Kennedy Library; Department of State, "Background Paper: The President's Visit to Venezuela and Colombia," December 1961, NSF: Trips and Conferences, Box 235; Teodoro Moscoso to Dungan, March 5, 1962, NSF: CO: Venezuela, General (2/62–3/62), Box 192; Agency for International Development, *Overseas Loans and Grants*, pp. 61, 185.
21. Joint embassy/AID message to State Department, August 20, 1962, NSF: CO: Venezuela, General (8/20/62–8/31/62), Box 192.
22. Oral History of William Gaud, pp. 20, 40, Johnson Library; Oral History of Sol Linowitz, ambassador to the Organization of American States, pp. 63–65, Johnson Library; Oral History of Covey T. Oliver, assistant secretary of state, Part II, pp. 8–9, Johnson Library.
23. Senate, Subcommittee on Multinational Corporations, *Multinational Corporations in Brazil and Mexico*, pp. 34–37.
24. Stewart to State Department, August 12, 1962, NSF: CO: Venezuela, General (7/1/62–8/29/62), Box 192.
25. Agency for International Development, *Overseas Loans and Grants*, p. 61; Liss, *Diplomacy and Dependency*, pp. 200–203; memorandum for Kenneth O'Donnell, White House, from Brubeck, August 28, 1962, NSF: CO: Venezuela, General (8/20/62–8/31/62), Box 192; Stewart to

State Department, August 12, 1962, NSF: CO: Venezuela, General (7/1–8/19/62), Box 192.

26. Memorandum of conversation between Kennedy and Ambassador Mayobre by Edwin Martin, assistant secretary of state, April 11, 1962, NSF: CO: Venezuela, General (4/62), Box 192.
27. Betancourt quoted in *Venezuela* 10 (May–June 1961): 3–4; *Libro amarillo, 1960,* pp. J–L, 69–71; Liss, *Diplomacy and Dependency,* pp. 183–189; Moscoso to State Department, August 29, 1961, NSF: CO: Venezuela, General (7/61–9/61), Box 192.
28. Lieuwen, *Generals,* pp. 114–121; Theodore C. Sorenson, *Kennedy,* pp. 535–537; Schlesinger, *Thousand Days,* p. 769; Rusk quoted in his telegram to the embassy, Caracas, April 17, 1962, NSF: CO: Venezuela, General (4/62), Box 192; *Libro amarillo, 1963,* p. xcvi.
29. Phyllis R. Parker, *Brazil and the Quiet Intervention, 1964.*
30. Lieuwen, *Generals,* p. 87; Liss, *Diplomacy and Dependency,* p. 202; memorandum for O'Donnell from Brubeck, August 28, 1962, NSF: CO: Venezuela, General (8/20/62–8/31/62), Box 192.
31. Central Intelligence Agency, "Survey of Latin America," April 1, 1964, National Security Council File on Latin America, Box 1, Johnson Library; Johnson to Leoni, April 14, 1967, CO 311 (Venezuela), Box 79, General Files, White House Central Files, Johnson Library; Oral History of Gaud, pp. 49–50, Kennedy Library; Stewart to State Department, June 13, 1962, and June 15, 1962, NSF: CO: Venezuela, General (6/5/62–6/30/62), Box 192.
32. U.S. Army, *Area Handbook for Venezuela,* pp. 292–303; Fred Fejes, "Public Policy in the Venezuelan Broadcasting Industry," *Inter-American Economic Affairs* 32 (Spring 1979): 9–10; Moscoso to Edward R. Murrow, director, United States Information Agency, July 5, 1961, Country File: Venezuela, Box 128, President's Office File, Kennedy Library; "Psychological Offensive in Latin America," paper by Berle, June 29, 1961, Berle Diary, Roosevelt Library. Berle suggested funding Latin American educators; all of his other suggestions in the paper were implemented.
33. Rusk to embassy, Caracas, August 20, 1962, NSF: CO: Venezuela, General (8/20/62–8/31/62), Box 192.
34. W. W. Rostow, "Guerilla Warfare in Underdeveloped Areas, in *The Viet-Nam Reader,* ed. Marcus Raskin and Bernard Fall, pp. 112–113.
35. AID, Caracas, to Schlesinger, September 21, 1962, Venezuela Folder, Box 23, Papers of Arthur M. Schlesinger, Jr., Kennedy Library.
36. Stewart to State Department, July 6, 1962, and Rusk to embassy, Caracas, July 9, 1962, NSF: CO: Venezuela, General (7/1–8/19/62), Box 192.
37. Stewart to State Department, May 10, 1962, NSF: CO: Venezuela, General (1/61–6/61), Box 192; Betancourt to Moscoso, February 3, 1962, Venezuela Folder, Box 23, Schlesinger Papers; *Venezuela* 11 (Spring 1963): 8.
38. Lieuwen, *Generals,* pp. 88–91; Levine, *Conflict and Change,* pp. 109–110.

39. Gott, *Guerilla Movements*, pp. 166–169; Levine, *Conflict and Political Change*, pp. 50–51; Powell, *Venezuelan Peasant*, pp. 94–95; Alexander, *Communist Party in Venezuela*, p. 73.
40. George Ball, acting secretary of state, to Johnson, April 15, 1964, CO 311 (Venezuela), Box 12, Confidential Files, White House Central Files, Johnson Library.
41. General Services Administration, *Public Papers of the President: Kennedy, 1963*, p. 243; "The President's Trip to Venezuela and Colombia," December 1961, NSF: Trips and Conferences, Box 235; Testimony of John A. McCone, director of CIA, in U.S. House, Committee on Foreign Affairs, Subcommittee on Inter-American Affairs, *Hearings, Castro-Communist Subversion in the Western Hemisphere*, 88th Cong., 1st sess., pp. 64–66.
42. For scholarly defenses of Betancourt's policies, see Levine, *Conflict and Change*, pp. 258–259; Salcedo-Bastardo, *Historia fundamental*, pp. 486–487; Alexander, *Democratic Revolution*. James F. Petras, *Politics and Social Structure in Latin America*, pp. 92–105, is critical of Betancourt's moderation.
43. Stewart to State Department, May 5, 1961, NSF: CO: Venezuela, General (1/61–6/61), Box 192; *DSB*, May 29, 1961, p. 821; International Bank for Reconstruction and Development, *The Economic Development of Venezuela*, pp. 34–46.
44. Stewart testimony in House, Committee on Foreign Affairs, *Castro-Communist Subversion*, p. 58; Stewart to State Department, June 13, 1962, NSF: CO: Venezuela, General (6/5/62–6/30/62), Box 192.
45. Stewart to State Department, August 12, 1962, NSF: CO: Venezuela, General (7/1–8/19/62), Box 192.
46. Jonathan V. Van Cleve, "The Latin American Policy of President Kennedy: A Reexamination Case: Peru," *Inter-American Economic Affairs* 30 (Spring 1977): 29–44.
47. Berle to Kennedy, "Report on the Task Force on Latin America," July 7, 1961, Box 220, Berle Diary; Levinson and de Onís, *Alliance*, pp. 52–58; Schlesinger, *Thousand Days*, p. 201.
4. Oral History of Chester Bowles, under secretary of state, pp. 30, 80, Kennedy Library; see also Ruth Leacock, "JFK, Business, and Brazil," *Hispanic American Historical Review* 59 (November 1979): 636–673.
49. Mann quoted in letter, Berle to Rusk, July 19, 1961, Box 220, Berle Diary; see also Mann's circular to all ambassadors to Latin America on the military assistance program, January 2, 1965, Country File: Latin America, National Security Council Files, Box 2, Johnson Library.
50. Johnson quoted in *DSB*, March 4, 1968, p. 325; Agency for International Development, *Overseas Loans and Grants*, p. 62; U.S. Department of Commerce, Bureau of International Commerce, *Basic Data on the Economy of Venezuela*, pp. 5, 21–23.
51. Oral History of Maurice M. Bernbaum, p. 23, Johnson Library.
52. For assessments of post-Betancourt Venezuela, see CIA, "Survey of Latin America," April 1, 1964, Country File: Latin America, Box 1, Na-

tional Security Council Files, Johnson Library; director, CIA, to White House, "National Intelligence Estimate: Venezuela," December 16, 1965, CO 311 (Venezuela), Box 12, Confidential Files, White House Central Files, Johnson Library.

53. *Libro amarillo, 1965*, pp. cxxv–cxxxiv, and *1968*, pp. cvii–cxii, 73–92; F. Parkinson, *Latin America, the Cold War, and the World Powers, 1945–1973*, pp. 186–189; *DSB*, August 10, 1964, pp. 174–184, and October 16, 1967, pp. 490–498.
54. Alexander, *Communist Party in Venezuela*, p. 202; D. Bruce Jackson, *Castro, the Kremlin, and Communism in Latin America*, pp. 5–7, 16.
55. Speech by Betancourt in *Venezuela* 10 (Winter 1961–1961): 2; Levinson and de Onís, *Alliance*, p. 60; *DSB*, December 31, 1962, p. 993.
56. Agency for International Development, *Overseas Loans and grants*, p. 61; Johnson to Leoni, April 4, 1967, CO 311 (Venezuela), General File, White House Central File, Johnson Library; Barry M. Blechman and Stephen S. Kaplan, *Force without War: United States Armed Forces as a Political Instrument*, pp. 46, 551–552; director, CIA, to White House, "National Intelligence Estimate: Venezuela," CO 311 (Venezuela), Box 12, Confidential Files, White House Central Files, Johnson Library.
57. Gott, *Guerilla Movements*, pp. 208–209; Levine, *Conflict and Political Change*, pp. 197–201, 207–208.
58. Rostow quoted in Stephen E. Ambrose, *Rise to Globalism: American Foreign Policy, 1938–1976*, p. 303.
59. Jackson, *Castro, the Kremlin*, pp. 44–46, 61, 93.
60. Alexander, *Communist Party in Venezuela*, p. 112, 198–204; Gott, *Guerilla Movements*, pp. 208–209, 215; House, Committee on Foreign Affairs, *Castro-Communist Subversion*, pp. 64–66.
61. Special memorandum No. 31-65 by Sherman Kent, chairman, Office of National Estimates, CIA, "Some Thoughts about the Latin American Left," December 29, 1965, Country File: Latin America, Box 2, National Security Files, Johnson Library; Rusk quoted in *DSB*, October 16, 1967, p. 491.
62. Speech by Betancourt to National Press Club, Washington, D.C., February 20, 1963, *Venezuela* 11 (Spring 1963): 10; Raúl Leoni, "View from Caracas," *Foreign Affairs* 43 (July 1965): 639–641.
63. *Venezuela* 9 (February 1959): 5; Tugwell, *Politics of Oil*, pp. 51–52.
64. *OGJ*, December 29, 1958, pp. 83–84.
65. Statement of Ambassador to United States Marcos Falcón-Briceño in *OGJ*, January 5, 1959, p. 88.
66. Creole Petroleum, *Data on Petroleum, 1974*, pp. 14, 18; Tugwell, *Politics of Oil*, pp. 79, 180–181.
67. Tugwell, *Politics of Oil*, p. 80; Lowry quoted in telegram #66, Bernbaum to State Department, June 6, 1966, File FI 11, Confidential Files, White House Central Files, Johnson Library; *OGJ*, October 26, 1959, pp. 46–47, August 15, 1960, p. 109, and August 15, 1966, pp. 39–41.
68. *OGJ*, January 5, 1959, p. 88.

69. Betancourt, *Venezuela*, pp. 393–394; interview with Juan Pablo Pérez Alfonzo, minister of mines and hydrocarbons, in *OGJ*, June 19, 1961, pp. 81–82.
70. Betancourt, *Venezuela*, p. 393.
71. Tugwell, *Politics of Oil*, p. 183; *OGJ*, October 17, 1966, pp. 72–81.
72. Humberto Peñaloza, "The Political Framework of Venezuelan Oil: Changes and Opportunities," in *Venezuela: 1969*, ed. Taylor, pp. 197–215.
73. John H. Lichtblau, "United States Oil Import Policies and Venezuelan Petroleum Exports" in *Venezuela: 1969*, ed. Taylor, pp. 182–193; *DSB*, November 13, 1967, pp. 638–642; Creole Petroleum, *Data on Petroleum, 1965*, p. 27, and *Data on Petroleum, 1974*, p. 27.
74. Eloy Porras, *Juan Pablo Pérez Alfonzo: El hombre que sacudió al mundo*, pp. 19–133; William Smith, "Unlikely Father of Arab Power," *New York Times*, December 12, 1973, p. 5; *Venezuela* 10 (November–December 1960): 5; Juan Pablo Pérez Alfonzo, *Petróleo y dependencia*, pp. 81–89.
75. Pérez Alfonzo, *Petróleo y dependencia*, pp. 71–74; idem, *Petróleo: Jugo de la tierra*, p. 41; Tugwell, *Politics of Oil*, pp. 32–37.
76. Werner Baer, "The Economics of Prebisch and ECLA," in *Latin America: Problems of Economic Development*, ed. Charles T. Nisbet, pp. 203–218; Pérez Alfonzo, *El pentágono*, pp. 83–88.
77. Ministerio de Minas e Hidrocarburos, *Memoria, 1960*, pp. 395–400 (hereafter cited as *Memoria* with appropriate year); Tugwell, *Politics of Oil*, pp. 54–60; *OGJ*, February 8, 1963, p. 74; quotation is from "Memorandum on Venezuelan Oil," July 22, 1959, National Security Council: Oil Folder, Box 14, Papers of Philip E. Areeda, special counsel to president, Eisenhower Library.
78. Pérez Alfonzo, *El pentágono*, pp. 59–64; Fuad Rouhani, *A History of O.P.E.C.*, pp. 75–82; Tariki quoted in *OGJ*, May 9, 1960, p. 99; *OGJ*, September 26, 1960, p. 120.
79. *Memoria, 1966*, pp. 22–23; Kim Faud, "Venezuela's Role in OPEC: Past, Present, and Future," in *Contemporary Venezuela and Its Role in International Affairs*, ed. Robert D. Bond, pp. 126–133; *OGJ*, March 26, 1962, pp. 102–103, and Janauary 4, 1965, pp. 72–73; Anthony Sampson, *The Seven Sisters: The Great Oil Companies and the World They Shaped*, pp. 156–184.
80. James Aikins, "The Oil Crisis: This Time the Wolf is Here," *Foreign Affairs* 51 (April 1973): 486; interview with Walter Levy, international oil consultant, in *OGJ*, June 28, 1965, pp. 44–46; *OGJ*, September 13, 1965, p. 76. I could not find any mention of OPEC in the speeches and policy statements reprinted in the *Department of State Bulletin* in the 1960s.
81. Pérez Alfonzo quoted in *OGJ*, June 19, 1960, pp. 80–81; Pérez Alfonzo, *Petróleo y dependencia*, pp. 241–242; *Memoria, 1964*, pp.16–17.
82. *Venezuela* 9 (June 1959): 4, and 11 (Spring 1963): 13–15; Pérez Alfonzo, *El pentágono*, pp. 29–34.

83. Proposal outlined in "Memorandum on Venezuelan Oil," July 22, 1959, National Security Council: Oil Folder, Box 12, Areeda Papers.
84. Memorandum of conversation with Representative Silvio Conte of Massachusetts by Presidential Assistant Allan Wallis, May 27, 1959, Tariff on Oil File (GF-142-F-1), White House General Files, Eisenhower Library.
85. Douglas R. Bohi and Milton Russell, *Limiting Oil Imports: An Economic History and Analysis*, pp. 144–161; Lichtblau, "United States Oil Import Policies," pp. 185–186, 190–191.
86. Department of State to Eisenhower, [n.d., but draft of memorandum is dated September 5, 1959], Oil Folder, Box 7, Papers of Don Paarlberg, special assistant to the president, Eisenhower Library.
87. Ibid.; "Memorandum on Venezuelan Oil," July 22, 1959, National Security Council: Oil Folder, Box 14, Areeda Papers; see also "Report of Myer Feldman on Discussions with President Betancourt on the United States Oil Import Program," December 29–30, 1962, President's Office Files: Subjects, Oil Imports Folder, Box 104, Kennedy Library.
88. Charles Schultze to Lee White, White House, June 24, 1965, File EX (TA6/Oil), Box 19, White House Central Files, Johnson Library. The executive branch had the authority to amend or abolish the quotas, since they were based on a national security finding of the Trade Agreements Extension Act.
89. Memorandum of conversation with Boyle by Walter I. Poyen, assistant to secretary of interior, March 9, 1964, File EX (TA6/Oil), White House Central File, Johnson Library; for position of independent oil producers, see Frank N. Ikard, president of American Petroleum Institute, to Johnson, November 7, 1968, File EX (TA6/Oil), White House Central Files, Johnson Library; for position of those opposed to quotas, see Gardner Ackley, chairman of Council of Economic Advisors, to Presidential Assistant Joseph Califano, Jr., January 14, 1966, CO 311 (Venezuela), Box 79, General Files, White House Central Files, Johnson Library.
90. Views of Standard Oil and Gulf in "Memorandum on Venezuelan Oil," July 22, 1959, National Security Council Folder, Box 14, Areeda Papers; Stewart to State Department, December 26, 1962, NSF: CO: Venezuela, General (12/62), Box 192; *OGJ*, July 20, 1959, p. 41; *The Lamp* 44 (Winter 1962): 2–5, and 47 (Summer 1965): 1.
91. Oral History of Bernbaum, pp. 19–21, Johnson Library; Special Assistant Lee White to Johnson, June 4, 1965, File EX (TA6/Oil), Box 19, White House Central Files, Johnson Library.
92. Betancourt, citing his conversations with Kennedy in February 1963, believes that if Kennedy had lived he would have granted Venezuelan petroleum parity with Canadian oil. See Betancourt, *Venezuela*, p. 391; Norman Gall, "The Challenge of Venezuelan Oil," *Foreign Policy* 18 (Spring 1975): 45.
93. State Department to Marvin Watson, White House, February 1, 1966, CO 311 (Venezuela), Box 12, Confidential Files, White House Central Files, Johnson Library; *Venezuela* 12 (Spring–Summer 1966): 5–6;

Leoni to Johnson, January 17, 1966, CO 311 (Venezuela), Box 12, Confidential Files, White House Central Files, Johnson Library; *OGJ*, February 28, 1966, p. 47.

94. Johnson to Leoni, August 8, 1967, CO 311 (Venezuela), Box 12, Confidential Files, White House Central Files, Johnson Library.
95. Oral History of Bernbaum, p. 19, Johnson Library.
96. *Memoria, 1967*, pp. 23–27; Betancourt, *Venezuela*, pp. 396–403; *Venezuela* 11 (Spring 1968): 2; Johnson to Leoni, August 8, 1967, CO 311 (Venezuela), Box 12, Confidential Files, White House Central Files, Johnson Library; *OGJ*, August 7, 1967, p. 96.
97. Betancourt quoted in *Venezuela* 11 (Winter 1963–1964): 5, and 11 (Summer 1963): 3.
98. Leoni to Johnson, January 17, 1966, and Johnson to Leoni, February 12, 1968, CO 311 (Venezuela), Box 12, Confidential Files, White House Central Files, Johnson Library; Leoni, "View from Caracas," p. 643; Oral History of Bernbaum, pp. 38–39, Johnson Library; speeches by Leoni in *Venezuela* 12 (Winter 1965–1966): 3, and 12 (Spring 1967): 2.
99. *OGJ*, August 16, 1968, p. 54; Tugwell, *Politics of Oil*, pp. 86–99; *Memoria, 1966*, pp. 8–10, and *1968*, pp. 4–7.
100. Tugwell, *Politics of Oil*, p. 167; Creole Petroleum, *Data on Petroleum, 1974*, p. 12.
101. For statistics, see Banco Central, *La economía venezolana*, pp. 54–56, 63; Baily, *Development of South America*, p. 6; Levinson and de Onis, *Alliance*, pp. 17, 230–233, 304–305; Robert J. Alexander, *Agrarian Reform in Latin America*, pp. 31–32; Commerce, Bureau of International Commerce, *Basic Data*, No. 71-058, pp. 3–17.
102. Leoni, "View from Caracas," p. 643.
103. Banco Central, *La economía venezolana*, pp. 72, 93; Baloyra, "Oil Policies," pp. 36–39.
104. Commerce, Bureau of International Commerce, *Basic Data*, No. 71-058, pp. 22–23; U.S. Department of Agriculture, *A Market for U.S. Products: Venezuela*, pp. 2–4; Creole Petroleum, *Data on Petroleum, 1974*, p. 25.
105. Creole Petroleum, *Data on Petroleum, 1974*, pp. 22–23; Senate, Subcommittee on Multinational Corporations, *Multinational Corporations in Brazil and Mexico*, pp. 34–37; *Memoria, 1974*, p. X-25, and *1975*, p. IX-454. The oil companies disputed the government's accounting of profits. They held, for example, that their return on net fixed assets in 1968 was only 27.75 percent; see Tugwell, *Politics of Oil*, pp. 180–181; see also Appendix, Tables A and B.

8. The Energy Crisis, 1969–1976

1. Salcedo-Bastardo, *Historia fundamental*, p. 487.
2. Tugwell, "Christian Democrats," pp. 245–267; *Venezuela* 13 (Fall 1970): 9.
3. Tugwell, *Politics of Oil*, pp. 100–103.
4. *Libro amarillo, 1972*, p. 6; Charles D. Ameringer, "The Foreign Policy

of Venezuelan Democracy," in *Venezuela: The Democratic Experience*, ed. John D. Martz and David J. Meyers, pp. 335–358; Winfield Burggraaff, "Oil and Caribbean Influence: The Role of Venezuela," in *Restless Caribbean: Changing Patterns of International Relations*, ed. Richard Millett, pp. 193–195.

5. For Caldera's speeches, see *Venezuela* 12 (Spring 1969): 10, 13 (Summer 1971): 6–7, and 12 (Fall 1970): 2–12.
6. For information on Third World economic issues, see Richard N. Cooper (ed.), *A Reordered World: Emerging International Economic Problems*; Joan Edelman Spero, *The Politics of International Economic Relations*; Robert W. Tucker, *The Inequality of Nations*.
7. Nixon quoted in *DSB*, November 17, 1969, pp. 409–414; *The Rockefeller Report on the Americas*; Council on Foreign Relations, William Lineberry, *The United States in World Affairs, 1970*, pp. 184–190.
8. Baily, *Development of South America*, pp. 120–121.
9. Ibid., pp. 122–125; *DSB*, November 17, 1969, pp. 409–414; Agency for International Development, *Overseas Loans and Grants*, p. 33.
10. *New York Times*, June 2, 1969, p. 9.
11. Church quoted in ibid., August 8, 1969, p. 7; Tugwell, *Politics of Oil*, p. 134; *OGJ*, August 18, 1969, p. 58.
12. *OGJ*, July 21, 1969, pp. 32–33.
13. U.S. Cabinet Task Force on Oil Import Control, *The Oil Import Question: A Report on the Relationship of Oil Imports to the National Security*, pp. 21–22, 38–41, 109; *OGJ*, July 21, 1969, pp. 32–33; *The Lamp* 52 (Fall 1970): 2; Lichtblau, "United States Oil Import Policies," pp. 182–193.
14. Sosa-Rodríguez quoted in *Venezuela* 13 (Winter 1969): 4–5, and 13 (Spring 1970): 4; Caldera quoted in Ameringer, "The Foreign Policy of Venezuelan Democracy," pp. 347–348; Tugwell, *Politics of Oil*, pp. 133–134.
15. Cabinet Task Force, *Oil Import Question*, pp. 34–37, 97–98, 109; *OGJ*, December 22, 1969, pp. 24–25.
16. *Venezuela* 13 (Fall 1970): 6–9, and 13 (Summer 1971): 6–7.
17. Cabinet Task Force, *Oil Import Question*, pp. 95–98; *OGJ*, December 22, 1969, pp. 24–25.
18. Bohi and Russell, *Limiting Oil Imports*, pp. 200–203; *OGJ*, August 24, 1970, pp. 30–31.
19. Caldera's speeches in the United States are reprinted in *Venezuela* 13 (Fall 1970): 2–9.
20. *OGJ*, June 8, 1970, p. 72; *Venezuela* 13 (Spring 1971): 7, and 13 (Summer 1971): 3.
21. Tugwell, *Politics of Oil*, pp. 108–129; Banco Central, *La economía venezolana*, pp. 54, 56.
22. *OGJ*, December 21, 1970, p. 34, December 28, 1970, p. 87, and January 3, 1972, pp. 21–22.
23. Ibid., August 9, 1971, p. 46; *New York Times*, June 18, 1971, pp. 1, 62, and June 23, 1971, pp. 63, 66.

24. McClintock's speech in *DSB*, April 19, 1971, pp. 522–528.
25. Undisclosed source quoted in Tugwell, *Politics of Oil*, p. 135.
26. *New York Times*, June 9, 1972, p. 36, and June 13, 1972, p. 2.
27. Tugwell, *Politics of Oil*, pp. 138–142; *Dallas Times Herald*, November 27, 1979, pp. 1, 10.
28. Statement by John Irwin, under secretary of state, to House Committee on Interior and Insular Affairs, *DSB*, May 1, 1972, pp. 626–631; Aikins, "The Oil Crisis," p. 489; *OGJ*, January 1, 1973, p. 21; *Latin America*, March 23, 1973, p. 92; *Venezuela* 14 (Summer 1973): 5; *OGJ*, July 16, 1973, p. 71.
29. *Latin America*, March 30, 1973, p. 102, and May 25, 1973, pp. 162–164; Francis Herron, "Venezuela, Its Oil, and the United States," *Hartford Courant*, November 10, 1973, p. 16; Tugwell, *Politics of Oil*, p. 140.
30. U.S. Senate, Subcommittee on Multinational Corporations of the Committee on Foreign Relations, *Report, Multinational Corporations and U.S. Foreign Policy*, 94th Cong., 1st sess., p. 14.
31. Cabinet Task Force, *Oil Import Question*, pp. 21–22; for statistics, see Bohi and Russell, *Limiting Oil Imports*, pp. 22–24, and Appendix, Table C.
32. Haider quoted in *The Lamp* 50 (Winter 1968): 1; see also *OGJ*, April 25, 1966, p. 97.
33. Aikins, "The Oil Crisis," pp. 463–465; Bohi and Russell, *Limiting Oil Imports*, pp. 203–205; *The Lamp* 56 (Summer 1974), 1. Prediction cited excludes production of natural gas liquids.
34. Cabinet Task Force, *Oil Import Question*, pp. 46–48, 68; Bohi and Russell, *Limiting Oil Imports*, p. 24.
35. Bohi and Russell, *Limiting Oil Imports*, pp. 8–9, 22–23; *OGJ*, August 2, 1971, pp. 32–33; Creole Petroleum, *Data on Petroleum, 1974*, p. 34.
36. Creole Petroleum, *Data on Petroleum*, p. 31. In 1973, the United States had 41.8 billion barrels of reserve. Total world reserves were 634.7 billion barrels, with 349.7 billion of those barrels in the Middle East.
37. The analyst quoted is Walter Levy, a consultant to oil companies, the United States, and the World Bank. *OGJ*, June 28, 1965, pp. 44–46.
38. Subcommittee on Multinational Corporations, *Report, Multinational Oil Corporations*, pp. 14–18; Faud, "Venezuela's Role in OPEC," pp. 133–136; *OGJ*, December 30, 1968, pp. 100–101; Tugwell, *Politics of Oil*, pp. 111–112, 129–131.
39. Subcommittee on Multinational Corporations, *Report, Multinational Oil Corporations*, pp. 121–150; Sampson, *Seven Sisters* pp. 208–229; *OGJ*, February 9, 1970, p. 29, and October 16, 1972, p. 78; *Latin America*, September 7, 1973, p. 283; *Memoria, 1973*, pp. 31–36, 51–63.
40. Aikins, "The Oil Crisis," pp. 462–490; *OGJ*, August 2, 1971, pp. 32–33, and May 15, 1972, p. 50; *DSB*, May 1, 1972, pp. 626–631; Subcommittee on Multinational Corporations, *Report, Multinational Oil Corporations*, p. 14.

41. Aikins in *OGJ*, May 15, 1972, p. 50; Irwin quoted in *DSB*, May 1, 1972, pp. 626–631.
42. Subcommittee on Multinational Corporations, *Report, Multinational Corporations*, p. 18.
43. Pérez de la Salvia quoted in *Memoria, 1972*, p. 2.
44. Tugwell, *Politics of Oil*, pp. 129–131; *Venezuela* 16 (November 1975): 15–18; *OGJ*, December 21, 1970, p. 34, and January 3, 1972, pp. 21–22.
45. Article by William D. Smith in *New York Times*, December 12, 1973, p. 5.
46. Caldera quoted in Tugwell, *Politics of Oil*, pp. 141–142; embassy quoted in *Venezuela* 15 (Winter 1973–1974): 11; *Memoria, 1973*, pp. 5–6; *New York Times*, April 27, 1975, III, p. 2.
47. Salcedo-Bastardo, *Historia fundmental*, pp. 487–488; *Venezuela* 15 (Winter 1973–1974): 3.
48. *Memoria, 1973*, p. 8; *Latin America*, December 25, 1973, p. 409; *OGJ*, December 31, 1973, p. 58.
49. Pérez quoted in *Latin America*, June 14, 1974, p. 181, and *Venezuela* 15 (Spring–Summer 1974): *New York Times*, March 12, 1974, p. 3; Pérez' letters reprinted in *Venezuela* 15 (November 1974): 8–9, 19, and 15 (December 1974): 16–17; Ambassador Miguel Angel Burelli's statement is in *Venezuela* 15 (November 1974): 6–7.
50. *Latin America*, April 5, 1974, pp. 108–109. The official quoted is Secretary of the Treasury George Shultz.
51. Ibid., January 10, 1975, p. 9.
52. See, for example, Ford's speech on September 18, 1974, to the United Nations in *DSB*, October 7, 1974, pp. 465–468.
53. Baily, *Development of South America*, pp. 125–128; speeches by Kissinger in *DSB*, March 18, 1974, pp. 257–264, March 24, 1975, pp. 361–369, and June 2, 1975, pp. 713–719.
54. Pérez quoted in *Latin America*, April 25, 1975, pp. 124–125.
55. Mary Jeanne Reid Martz, "SELA: The Latin American Economic System, 'Ploughing the Seas'?" *Inter-American Economic Affairs* 32 (Spring 1979): 47–48; *Latin America*, December 3, 1976, pp. 374–375.
56. James F. Petras, *Critical Perspectives on Imperialism and Social Class in the Third World*, p. 246; *Latin America*, February 28, 1975, pp. 65–66, and September 26, 1975, pp. 298, 300.
57. *Memoria, 1974*, pp. 30–33, and *1975*, pp. 2–8; Pérez quoted in *OGJ*, July 14, 1975, p. 45; *OGJ*, January 12, 1976, p. 36; Petras, *Critical Perspectives*, pp. 229–246.
58. *OGJ*, November 3, 1975, p. 19; Petras, *Critical Perspectives*, pp. 229, 245.
59. *OGJ*, February 25, 1974, p. 29; *New York Times*, November 3, 1975, pp. 57, 59; Petras, *Critical Perspectives*, p. 229; *Latin America Weekly Report*, January 18, 1980, p. 5.
60. *DSB*, December 30, 1974, p. 918.
61. *New York Times*, June 30, 1975, pp. 43, 45; *Latin America*, August 30,

1974, pp. 265–266; *Venezuela* 16 (November 1975): 15–18; J. K. Jamieson of Exxon quoted in *The Lamp* 56 (Fall 1974): 1.

62. *DSB*, March 15, 1976, pp. 313–321, 327–331; *Venezuela* 17 (April 1976): 7.
63. World Bank, *World Bank Atlas, 1977*, p. 6; *Venezuela* 17 (April 1976): 8; Burggraaff, "Oil and Caribbean Influence," p. 195.
64. Salcedo-Bastardo, *Historia fundamental*, pp. 519–521, 529–533.
65. Ibid., pp. 542–543, 545; Baily, *Development of South America*, p. 6.
66. Betancourt, *Venezuela's Oil*, pp. 262–265. See also essay, "La pérdida de nuestra identidad nacional," by Domingo A. Labarca Prieto in *El Nacional*, June 30, 1980, p. A4.
67. *Libro amarillo, 1973*, pp. P–Q, 331–347; Martz, "SELA," pp. 36–41; Ameringer, "The Foreign Policy of Venezuelan Democracy," pp. 355–356.
68. Quotation from *New York Times*, May 11, 1975, IV, p. 2; Martz, "SELA," p. 55; Fejes, "Venezuelan Broadcasting Industry," pp. 20–24; John V. Lombardi, "The Patterns of Venezuela's Past," in Martz and Meyers (eds.), *Venezuela*, p. 19.
69. U.S. Department of Commerce, Bureau of International Commerce, *Economic Trends and Their Implications for the United States: Venezuela*, Report No. 93, pp. 1–21, and Report No. 54, pp. 3–12; idem, *Venezuela: A Survey of U.S. Business Opportunities*; U.S. Department of State, Office of Public Communications, Bureau of Public Affairs, *The Trade Debate*, p. 13; see also Appendix, Table A.
70. Subcommittee on Multinational Corporations, *Multinational Corporations in Brazil and Mexico*, pp. 34–37; U.S. Department of State, *Background Notes: Venezuela*, p. 6; *Latin America Weekly Report*, November 16, 1979, pp. 32–33; Burggraaff, "Oil and Caribbean Influence," pp. 200–201; Gall, "The Challenge of Venezuelan Oil," pp. 58–60; see also Appendix, Table B.
71. Commerce, *Economic Trends: Venezuela*, Report No. 54, pp. 6–8; *OGJ*, January 30, 1978, pp. 131, 136; U.S. Arms Control and Disarmament Agency, *World Military Expenditures and Arms Transfers, 1967–1976*, p. 160; see also Appendix, Table D.
72. Burggraaff, "Oil and Caribbean Influence," pp. 195–200; *Venezuela* 16 (January 1975): 4–7, 10; Liss, *Diplomacy and Dependency*, p. 324.
73. Quoted in Gall, "The Challenge of Venezuelan Oil," p. 52; John Martz, "Venezuelan Foreign Policy Toward Latin America," in Bond (ed)., *Contemporary Venezuela*, pp. 187–195.
74. See President Jimmy Carter's speech, "A Just International Order," to the Venezuelan Congress on March 29, 1978, in *DSB* 77 (May 1978): 1–4.
75. Pérez quoted in *Venezuela* 17 (January 1976): 3.
76. The scholar quoted is Sheldon Liss in *Diplomacy and Dependency*, p. 274; for an analysis of the Venezuelan electorate, see Enrique Baloyra and John D. Martz, *Political Attitudes in Venezuela*, p. xvii.

Selected Bibliography

NOTE ON SOURCES

The most valuable materials for analyzing U.S. policies toward Venezuela were the records of the Department of State, which are located in the National Archives in Washington, D.C. The *Papers Relating to the Foreign Relations of the United States* (1895–1951) and the *Department of State Bulletin* (1946–1978) supplemented the State Department's unpublished memorandums and despatches. When I was conducting my research, the files of the State Department were open only through 1949. But, by applying the Freedom of Information Act, I gained access to the Venezuelan files of the National Security Council for the Eisenhower, Kennedy, and Johnson administrations. Particularly during the 1960s, the National Security Council received copies of the correspondence between the State Department and its field officers in Venezuela.

In addition to holding the files of key executive agencies, such as the National Security Council, the Hoover, Roosevelt, Truman, Eisenhower, Kennedy, and Johnson presidential libraries own other rich research materials. Diplomats, businessmen, trade organizations, and foreign leaders often circumvented the Department of State and wrote directly to the president. The libraries' comprehensive subject, country, and name indexes facilitated research in the presidential papers. The presidential libraries also hold the private papers and oral histories of foreign service officers and public officials. Of the many manuscript collections consulted, the papers of Frank P. Corrigan, an ambassador to Venezuela, and Adolf A. Berle, an influential advisor to Presidents Roosevelt and Kennedy, proved especially valuable. The Corrigan and Berle papers are in the Roosevelt Library.

In view of the prominent role that Great Britain and British oil companies played in Venezuela, the files of the Foreign Office in the Public Record Office in London were essential for evaluating Venezuelan-U.S. relations. The Foreign Office's records not only were important for recounting the Anglo-American rivalry for Venezuelan oil but also were excellent supplements to U.S. materials. British diplomats were astute analysts of the Venezuelan political milieu, and their reports served as bases for assessing Washington's interpretations of Venezuela's policies. Another European government source that I used was the records of the German Foreign Ministry in the

National Archives Collection of Captured Foreign Documents. The German records provided information on Nazi activities in Venezuela.

Venezuela's twentieth-century governmental archives were closed to scholars during the time I worked on this study in Caracas. For the Venezuelan perspective, I relied on annual reports, *Memorias*, of such agencies as the Ministry of Fomento (Development) and Ministry of Hydrocarbons and the *Libro amarillo* published yearly by the Venezuelan Foreign Ministry. Since 1950, Venezuela's embassy in Washington has published a journal, *Venezuela Up-To-Date*, that reprints key addresses and policy statements by Venezuela officials on relations with the United States and the multinational oil companies. In compiling lists of books and articles on Venezuelan history and government, I found John V. Lombardi's *Venezuelan History: A Comprehensive Working Bibliography* to be invaluable.

The Oil and Gas Journal presented the industry's point of view. Other industry sources were the companies' official histories and their journals, such as Standard Oil of New Jersey's *The Lamp*. Two studies of the Venezuelan oil industry that I frequently cited were Edwin Lieuwen's *Petroleum in Venezuela: A History* and Franklin Tugwell's *The Politics of Oil in Venezuela*. Both are models of scholarship.

PRIMARY SOURCES

Manuscript Collections

Areeda, Philip E. Dwight D. Eisenhower Library. Abilene, Kansas.
Austin, Warren R. Bailey Library, University of Vermont. Burlington, Vermont.
Berle, Adolf A. Franklin D. Roosevelt Library. Hyde Park, New York.
Bohan, Merwin L. Harry S Truman Library. Independence, Missouri.
Byrnes, James F. Cooper Library, Clemson University. Clemson, South Carolina.
Corrigan, Frank P. Franklin D. Roosevelt Library. Hyde Park, New York.
Dulles, John Foster. Firestone Library, Princeton University. Princeton, New Jersey.
Feely, Edward F. Herbert Hoover Library. West Branch, Iowa.
Feis, Herbert. Library of Congress. Washington, D.C.
Forrestal, James V. Firestone Library, Princeton University. Princeton, New Jersey.
Grew, Joseph Clark. Houghton Library, Harvard University. Boston, Massachusetts.
Harrison, Leland. Library of Congress. Washington, D.C.
Hull, Cordell. Library of Congress. Washington, D.C.
Inman, Samuel Guy. Library of Congress. Washington, D.C.
Krug, Julius A. Library of Congress. Washington, D.C.
Leahy, William. Library of Congress. Washington, D.C.
Miller, Edward G., Jr. Harry S Truman Library. Independence, Missouri.
Morgenthau, Henry, Jr. Franklin D. Roosevelt Library. Hyde Park, New York.

Ogilvie-Forbes, Sir George. Kings College Library. Aberdeen, Scotland.
Paarlberg, Don. Dwight D. Eisenhower Library. Abilene, Kansas.
Phillips, William. Houghton Library, Harvard University. Boston, Massachusetts.
Schlesinger, Arthur M., Jr. John F. Kennedy Library. Boston, Massachusetts.
Sorenson, Theodore C. John F. Kennedy Library. Boston, Massachusetts.
Spingarn, Stephen J. Harry S Truman Library. Independence, Missouri.
Stimson, Henry L. Sterling Memorial Library, Yale University. New Haven, Connecticut.
Waugh, Samuel. Dwight D. Eisenhower Library. Abilene, Kansas.
White, Harry Dexter. Firestone Library, Princeton University. Princeton, New Jersey.

Presidential Papers

Coolidge, Calvin. Library of Congress. Washington, D.C.
Eisenhower, Dwight D. Dwight D. Eisenhower Library. Abilene, Kansas.
Hoover, Herbert. Herbert Hoover Library. West Branch, Iowa.
Johnson, Lyndon Baines. Lyndon Baines Johnson Library. Austin, Texas.
Kennedy, John F. John F. Kennedy Library. Boston, Massachusetts.
Roosevelt, Franklin D. Franklin D. Roosevelt Library. Hyde Park, New York.
Roosevelt, Theodore. Library of Congress. Washington, D.C.
Taft, William Howard. Library of Congress. Washington, D.C.
Truman, Harry S. Harry S Truman Library. Independence, Missouri.
Wilson, Woodrow. Library of Congress. Washington, D.C.

Unpublished Government Documents and Papers

Germany. Records of the Foreign Ministry. U.S. National Archives Collection of Captured Foreign Documents. Washington, D.C.
United Kingdom. Consular Office Correspondence. Public Record Office. London.
———. Foreign Office Correspondence. Public Record Office. London.
United States. General Records of the Department of State. Record Group 59, National Archives. Washington, D.C.
———. National Security Council Files, 1961–1963. John F. Kennedy Library. Boston, Massachusetts.
———. National Security Council Files, 1963–1969. Lyndon Baines Johnson Library. Austin, Texas.
———. Records of the Joint Army and Navy Advisory Board on American Republics. Modern Military Records. Record Group 225, National Archives. Washington, D.C.
———. Records of the Special Service Squadron. General Records of the Department of the Navy. Record Group 80, National Archives. Washington, D.C.

Oral Histories

Bernbaum, Maurice M. Lyndon Baines Johnson Library. Austin, Texas.
Bohan, Merwin L. Harry S Truman Library. Independence, Missouri.

Bowles, Chester. John F. Kennedy Library. Boston, Massachusetts.
Cabot, John M. John Foster Dulles Oral History Project. Firestone Library, Princeton University. Princeton, New Jersey.
Dillon, C. Douglas. John Foster Dulles Oral History Project. Firestone Library, Princeton University. Princeton, New Jersey.
Dreier, John C. John Foster Dulles Oral History Collection. Firestone Library, Princeton University. Princeton, New Jersey.
Eisenhower, Milton. Columbia University Oral History Project. New York, New York.
Figueres, José. Harry S Truman Library. Independence, Missouri.
Gaud, William. John F. Kennedy Library. Boston, Massachusetts.
———. Lyndon Baines Johnson Library. Austin, Texas.
Jamieson, Francis. Columbia University Oral History Project. New York, New York.
Linowitz, Sol. Lyndon Baines Johnson Library. Austin, Texas.
Mann, Thomas C. Lyndon Baines Johnson Library. Austin, Texas.
Oliver, Covey T. Lyndon Baines Johnson Library. Austin, Texas.
Rubottom, Roy Richard. John Foster Dulles Oral History Project. Firestone Library, Princeton University. Princeton, New Jersey.
White, John Campbell. Columbia University Oral History Project. New York, New York.

Published Government Documents and Papers

U.S. Agency for International Development. *U.S. Overseas Loans and Grants and Assistance from International Organizations: Obligations and Loan Authorizations, July 1, 1945–June 30, 1973*. Washington: Government Printing Office, 1974.
U.S. Arms Control and Disarmament Agency. *World Military Expenditures and Arms Transfers, 1967–1976*. Washington: Government Printing Office, 1978.
U.S. Army. *Area Handbook for Venezuela*. Washington: Government Printing Office, 1964.
U.S. Cabinet Task Force on Oil Import Control. *The Oil Import Question: A Report on the Relationship of Oil Imports to the National Security*. Washington: Government Printing Office, 1970.
U.S. Congress. *Congressional Record*. 63rd Cong., 1st sess., Vol. 50, Part 6, pp. 5845–5846.
———. House Committee on Foreign Affairs. Subcommittee on Inter-American Affairs. *Hearings, Castro-Communist Subversion in the Western Hemisphere*. 88th Cong., 1st sess. Washington: Government Printing Office, 1963.
———. ———. Special Subcommittee on Armed Services. *A Report of Investigation of Petroleum in Relation to National Defense*. 80th Cong., 2nd sess. Washington: Government Printing Office, 1948.
———. Senate. *Mutual Security Act of 1956*. Washington: Government Printing Office, 1956.
———. ———. Committee on Finance. *Hearings on H.R. 1 (Trade Agree-*

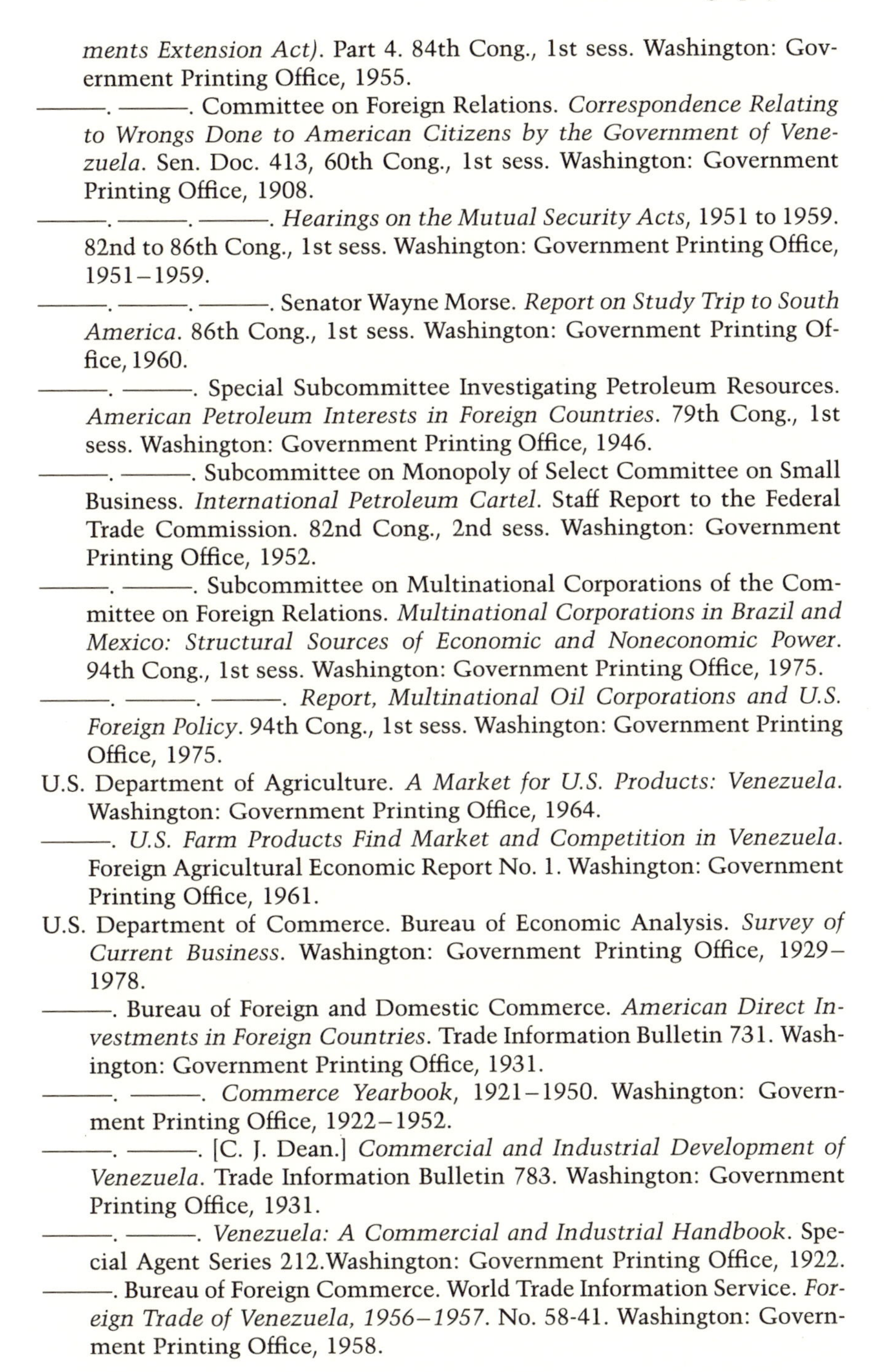

ments Extension Act). Part 4. 84th Cong., 1st sess. Washington: Government Printing Office, 1955.

———. ———. Committee on Foreign Relations. *Correspondence Relating to Wrongs Done to American Citizens by the Government of Venezuela*. Sen. Doc. 413, 60th Cong., 1st sess. Washington: Government Printing Office, 1908.

———. ———. ———. *Hearings on the Mutual Security Acts*, 1951 to 1959. 82nd to 86th Cong., 1st sess. Washington: Government Printing Office, 1951–1959.

———. ———. ———. Senator Wayne Morse. *Report on Study Trip to South America*. 86th Cong., 1st sess. Washington: Government Printing Office, 1960.

———. ———. Special Subcommittee Investigating Petroleum Resources. *American Petroleum Interests in Foreign Countries*. 79th Cong., 1st sess. Washington: Government Printing Office, 1946.

———. ———. Subcommittee on Monopoly of Select Committee on Small Business. *International Petroleum Cartel*. Staff Report to the Federal Trade Commission. 82nd Cong., 2nd sess. Washington: Government Printing Office, 1952.

———. ———. Subcommittee on Multinational Corporations of the Committee on Foreign Relations. *Multinational Corporations in Brazil and Mexico: Structural Sources of Economic and Noneconomic Power*. 94th Cong., 1st sess. Washington: Government Printing Office, 1975.

———. ———. ———. *Report, Multinational Oil Corporations and U.S. Foreign Policy*. 94th Cong., 1st sess. Washington: Government Printing Office, 1975.

U.S. Department of Agriculture. *A Market for U.S. Products: Venezuela*. Washington: Government Printing Office, 1964.

———. *U.S. Farm Products Find Market and Competition in Venezuela*. Foreign Agricultural Economic Report No. 1. Washington: Government Printing Office, 1961.

U.S. Department of Commerce. Bureau of Economic Analysis. *Survey of Current Business*. Washington: Government Printing Office, 1929–1978.

———. Bureau of Foreign and Domestic Commerce. *American Direct Investments in Foreign Countries*. Trade Information Bulletin 731. Washington: Government Printing Office, 1931.

———. ———. *Commerce Yearbook*, 1921–1950. Washington: Government Printing Office, 1922–1952.

———. ———. [C. J. Dean.] *Commercial and Industrial Development of Venezuela*. Trade Information Bulletin 783. Washington: Government Printing Office, 1931.

———. ———. *Venezuela: A Commercial and Industrial Handbook*. Special Agent Series 212.Washington: Government Printing Office, 1922.

———. Bureau of Foreign Commerce. World Trade Information Service. *Foreign Trade of Venezuela, 1956–1957*. No. 58-41. Washington: Government Printing Office, 1958.

———. Bureau of International Commerce. *Basic Data on the Economy of Venezuela*. Overseas Business Report No. 71-058. Washington: Government Printing Office, 1971.

———. ———. *Economic Trends and Their Implications for the United States: Venezuela*. Washington: Government Printing Office, 1968–1977.

———. ———. *Venezuela: A Survey of U.S. Business Opportunities*. Washington: Government Printing Office, 1976.

———. Office of Business Economics. *Business Statistics, 1961*. Washington: Government Printing Office, 1961.

———. Office of International Trade. *Foreign Commerce Yearbook*, 1948–1950. Washington: Government Printing Office, 1950–1952.

———. ———. *Investment in Venezuela: Conditions and Outlooks for United States Investors*. Washington: Government Printing Office, 1952.

U.S. Department of the Interior. Bureau of Mines. *Minerals Yearbook*. Washington: Government Printing Office, 1932–1978.

U.S. Department of State. *Allocation of Tariff Quota on Crude Petroleum and Fuel Oil*. Executive Agreements Series 191. Washington: Government Printing Office, 1941.

———. *Analysis of the Reciprocal Concessions and General Provisions of the Supplementary Trade Agreement between the United States and Venezuela, Signed at Caracas, Venezuela, August 18, 1952*. Washington: Government Printing Office, 1952.

———. *Background Notes: Venezuela*. Washington: Government Printing Office, 1979.

———. *Consultation among the American Republics with Respect to the Argentine Situation*. Washington: Government Printing Office, 1946.

———. *Department of State Bulletin*. Washington: Government Printing Office, 1946–1978.

———. *Papers Relating to the Foreign Relations of the United States*, 1895–1951. Washington: Government Printing Office, 1896–1980.

———. *Thirty-Third Report to the Congress on Lend-Lease Operations for the Period Ending December 31, 1951*. Washington: Government Printing Office, 1952.

———. Eleanor Lansing Dulles. *The Export Import Bank of Washington: The First Ten Years*. Commerical Policy Series No. 75. Washington: Government Printing Office, 1944.

———. Office of Public Affairs. *Venezuela: Oil Transforms a Nation*. Washington: Government Printing Office, 1953.

———. Office of Public Communications. Bureau of Public Affairs. *The Trade Debate*. Washington: Government Printing Office, 1978.

U.S. Federal Trade Commission. *The Report of the Federal Trade Commission on Foreign Ownership in the Petroleum Industry*. Washington: Government Printing Office, 1923.

U.S. General Services Administration. *Public Papers of the President*, Herbert Hoover to Gerald R. Ford, 1929–1976. 35 vols. Washington: Government Printing Office, 1958–1978.

U.S. Tariff Commission. *Commercial Policies and Trade Relations of the European Possessions in the Caribbean Area*. Washington: Government Printing Office, 1943.

———. *Foreign Trade of Latin America*. Section 10: *Venezuela*. Washington: Government Printing Office, 1940.

———. *Report to the Congress on the Costs of Crude Petroleum*. Washington: Government Printing Office, 1931.

———. *Report to the House of Representatives on the Cost of Production of Crude Petroleum*. Washington: Government Printing Office, 1932.

Venezuela. Banco Central. *La economía venezolana en los últimos trienta años*. Caracas: Imprenta Nacional, 1971.

———. Betancourt, Rómulo. *Venezuela and the U.S. Alliance for Progress*. Caracas: Imprenta Nacional, 1961.

———. Embassy of Venezuela, Washington, D.C. *Venezuela Up-To-Date*. Washington: Information Service of Embassy of Venezuela, 1950–1978.

———. Ministerio de Fomento. *Memoria*. Caracas: Imprenta Nacional, 1936–1950.

———. Ministerio de Minas e Hidrocarburos. *Memoria*. Caracas: Imprenta Nacional, 1960–1977.

———. Ministerio de Relaciones Exteriores. *El libro amarillo de los Estados Unidos de Venezuela*. Caracas, 1909–1977.

World Bank. *World Bank Atlas, 1977*. Washington: World Bank Press, 1977.

Autobiographies, Memoirs, Published Papers

Berle, Beatrice Bishop, and Jacobs, Travis Beale, eds. *Navigating the Rapids, 1918–1971: From the Papers of Adolf A. Berle*. New York: Harcourt Brace Jovanovitch, 1973.

Betancourt, Rómulo. *Venezuela: Oil and Politics*. Boston: Houghton Mifflin Co., 1979.

———. *Venezuela's Oil*. Boston: Allen & Unwin, 1978.

Braden, Spruille. *Diplomats and Demagogues: The Memoirs of Spruille Braden*. New Rochelle, N.Y.: Arlington House, 1971.

Campbell, Thomas M., and Herring, George C., eds. *The Diaries of Edward R. Stettinius, 1943–1946*. New York: New Viewpoints, 1975.

Gantenbein, James W., ed. *The Evolution of Our Latin American Policy: A Documentary Record*. New York: Columbia University Press, 1950.

Hoover, Herbert. *The Memoirs of Herbert Hoover*. Vol. II: *The Cabinet and Presidency*. New York: Macmillan Co., 1952.

Hull, Cordell. *The Memoirs of Cordell Hull*. 2 vols. New York: Macmillan Co., 1948.

López Contreras, Eleazar. *Gobierno y administración, 1936–1941*. Caracas: Editorial Arte, 1966.

Medina Angarita, Isaías. *Cuatro años de democracía*. Caracas: Pensamiento Vivo, 1963.

Morison, Elting, ed. *The Letters of Theodore Roosevelt*. 8 vols. Cambridge: Harvard University Press, 1951.
Nixon, Richard M. *Six Crises*. New York: Pyramid Books, 1962.
Pérez Jiménez, Marcos. *Diez años de desarrollo*. Caracas: Equipos Perezjimenistas y Desarrollistas, 1973.
Rosenman, Samuel I. *The Public Papers and Addresses of Franklin Delano Roosevelt*. 13 vols. New York: Russell, 1938–1950.
Schlesinger, Arthur M., Jr. *A Thousand Days: John F. Kennedy in the White House*. Boston: Houghton Mifflin Co., 1965.
Sorensen, Theodore C. *Kennedy*. New York: Harper & Row, 1965.
Vallenilla Lanz, Laureano. *Escrito de memoria*. Mexico City: Editorial Mazatlán, 1961.

Journals, Newspapers, Periodicals

The Lamp. New York, 1924–1976.
Latin America. London, 1972–1978.
Latin America Weekly Report. London, 1979–1980.
El Nacional. Caracas, 1979–1980.
New York Times. New York, 1969–1978.
Oil and Gas Journal. Tulsa, Oklahoma, 1922–1976.
World Petroleum. New York, 1933–1945.

SECONDARY SOURCES

Books

Acosta Saignes, Miguel. *Latifundio*. Mexico City: Editorial Popular, 1938.
Adams, Frederick C. *Economic Diplomacy: The Export-Import Bank and American Foreign Policy, 1934–1939*. Columbia: University of Missouri Press, 1976.
Alexander, Robert J. *Agrarian Reform in Latin America*. New York: Macmillan Co., 1974.
———. *The Communist Party in Venezuela*. Stanford: Stanford University Press, 1969.
———. *Latin American Political Parties*. New York: Praeger, 1973.
———. *The Venezuelan Democratic Revolution: A Profile of the Regime of Rómulo Betancourt*. New Brunswick: Rutgers University Press, 1964.
Ambrose, Stephen E. *Rise to Globalism: American Foreign Policy, 1938–1976*. New York: Penguin Books, 1976.
Arcaya, Pedro Manuel. *The Gómez Regime in Venezuela and Its Background*. Washington: Sun Printing Co., 1936.
Arnold, Ralph, et al. *The First Big Oil Hunt: Venezuela, 1911–1916*. New York: Vantage Press, 1960.
Bailey, Thomas A. *The Policy of the United States toward the Neutrals, 1917–1918*. Baltimore: Johns Hopkins University Press, 1942.
Baily, Samuel L. *The United States and the Development of South America, 1945–1975*. New York: New Viewpoints, 1976.

Baloyra, Enrique, and Martz, John D. *Political Attitudes in Venezuela*. Austin: University of Texas Press, 1979.

Baptista, Federico G. *Historia de la industria petrolera en Venezuela*. Caracas: Creole Petroleum Corporation, 1966.

Beale, Howard K. *Theodore Roosevelt and the Rise of America to World Power*. Baltimore: Johns Hopkins University Press, 1956.

Blechman, Barry M., and Kaplan, Stephen S. *Force without War: United States Armed Forces as a Political Instrument*. Washington: Brookings Institute, 1978.

Bohi, Douglas R., and Russell, Milton. *Limiting Oil Imports: An Economic History and Analysis*. Baltimore: Johns Hopkins University Press, 1978.

Bond, Robert D., ed. *Contemporary Venezuela and Its Role in International Affairs*. New York: New York University Press, 1978.

Brandes, Joseph. *Herbert Hoover and Economic Diplomacy: Department of Commerce Policy, 1921–1933*. Berkeley: University of California Press, 1963.

Burggraaff, Winfield J. *The Venezuelan Armed Forces in Politics, 1935–1959*. Columbia: University of Missouri Press, 1972.

Campbell, Charles S. *The Transformation of American Foreign Relations, 1965–1900*. New York: Harper & Row, 1976.

Clinton, Daniel Joseph. [Thomas Rourke.] *Gómez: Tyrant of the Andes*. New York: W. Morrow & Co., 1936.

Collier, Peter, and Horowitz, David. *The Rockefellers: An American Dynasty*. New York: Holt, Rinehart & Winston, 1976.

Conn, Stetson, and Fairchild, Byron. *The Framework of Hemisphere Defense*. Washington: Government Printing Office, 1960.

Cooper, Richard N., ed. *A Reordered World: Emerging International Economic Problems*. Washington: Potomac Associates, 1973.

Council on Foreign Relations et al. *The United States in World Affairs*, 1931–1970. New York: Simon & Schuster, 1932–1972.

Creole Petroleum Corporation. *Data on Petroleum and Economy of Venezuela, 1957–1974*. Caracas: Creole Petroleum Corporation, 1958–1975.

Cronon, E. David. *Josephus Daniels in Mexico*. Madison: University of Wisconsin Press, 1960.

DeConde, Alexander. *Herbert Hoover's Latin American Policy*. Stanford: Stanford University Press, 1951.

Denny, Ludwell. *We Fight for Oil*. New York: A. A. Knopf, 1928.

DeNovo, John A. *American Interests and Policies in the Middle East, 1900–1939*. Minneapolis: University of Minnesota Press, 1963.

Eisenhower, Milton. *The Wine is Bitter*. Garden City, N.Y.: Doubleday, 1963.

Everest, Allan S. *Morgenthau, the New Deal, and Silver*. New York: King's Crown Press of Columbia University, 1950.

Fernández, Pablo Emilio. *Gómez: El rehabilitador*. Caracas: J. Villegas, 1956.

Frye, Alton. *Nazi Germany and the American Hemisphere, 1933–1944.* New Haven: Yale University Press, 1967.

Furtado, Celso. *Economic Development of Latin America: A Survey from Colonial Times to the Cuban Revolution.* Cambridge: at the University Press, 1970.

Gallegos Ortíz, Rafael. *La cachorro: Juan Vicente Gómez.* Caracas: Editorial Fuentes, 1977.

———. *La historia política de Venezuela de Cipriano Castro a Pérez Jiménez.* Caracas: Imprenta Universitaria, 1960.

Gardner, Lloyd C. *Economic Aspects of New Deal Diplomacy.* Madison: University of Wisconsin Press, 1964.

Gellman, Irwin F. *Good Neighbor Diplomacy: United States Policies in Latin America, 1933–1945.* Baltimore: Johns Hopkins University Press, 1979.

Gerretson, Frederick C. *History of the Royal Dutch.* 4 vols. Leiden: Brill, 1958.

Gibb, George S., and Knowlton, Evelyn H. *History of Standard Oil (New Jersey).* Vol. II: *The Resurgent Years, 1911–1927.* New York: Harper & Row, 1956.

Giddens, Paul H. *Standard Oil Company (Indiana): Oil Pioneer of the Middle West.* New York: Appleton-Century-Crofts, 1955.

Gilmore, Robert L. *Caudillism and Militarism in Venezuela, 1810–1910.* Athens: Ohio University Press, 1964.

Gott, Richard. *Guerilla Movements in Latin America.* Garden City, N.Y.: Doubleday, 1971.

Green, David. *The Containment of Latin America.* Chicago: Quadrangle, 1971.

Guerrant, Edward O. *Roosevelt's Good Neighbor Policy.* Albuquerque: University of New Mexico Press, 1950.

Hogan, Michael J. *Informal Entente: The Private Structure of Cooperation in Anglo-American Economic Diplomacy, 1918–1928.* Columbia: University of Missouri Press, 1977.

Holt, W. Stull. *Treaties Defeated by the Senate.* Baltimore: Johns Hopkins University Press, 1933.

Hood, Miriam. *Gunboat Diplomacy, 1895–1905: Great Power Pressure in Venezuela.* Cranbury, N.J.: A. S. Barnes, 1977.

International Bank for Reconstruction and Development. *The Economic Development of Venezuela.* Baltimore: Johns Hopkins University Press, 1961.

Izard, Miguel, ed. *Política y economía en Venezuela, 1810–1976.* Caracas: Fundación John Boulton, 1976.

———. *Series estadísticas para la historia de Venezuela.* Mérida: Universidad de los Andes, 1970.

Jackson, D. Bruce. *Castro, the Kremlin, and Communism in Latin America.* Baltimore: Johns Hopkins University Press, 1969.

Jessup, Philip C. *Elihu Root.* 2 vols. New York: Dodd, Mead & Co., 1938.

Johnson, John. *Political Change in Latin America: The Growth of the Middle Sectors*. Stanford: Stanford Universtiy Press, 1958.

Kneer, Warren G. *Great Britain and the Caribbean, 1901–1913: A Study in Anglo-American Relations*. East Lansing: Michigan State University Press, 1975.

Kolb, Glen L. *Democracy and Dictatorship in Venezuela, 1945–1958*. New London, Conn.: Archon Books, 1974.

LaFeber, Walter. *The New Empire: An Interpretation of American Expansionism, 1860–1898*. Ithaca: Cornell University Press, 1963.

Larson, Henrietta M.; Knowlton, Evelyn H.; and Popple, Charles S. *History of Standard Oil Company (New Jersey)*. Vol III: *New Horizons, 1927–1950*. New York: Harper & Row, 1971.

Levine, Daniel H. *Conflict and Political Change in Venezuela*. Princeton: Princeton University Press, 1973.

Levinson, Jerome, and de Onís, Juan. *The Alliance That Lost Its Way: A Critical Report on the Alliance for Progress*. Chicago: Quadrangle, 1970.

Lewis, Cleona. *America's Stake in International Investments*. Washington: Brookings Institute, 1938.

Lieuwen, Edwin. *Generals vs. Presidents: Neo-Militarism in Latin America*. New York: Praeger, 1964.

———. *Petroleum in Venezuela: A History*. Berkeley: University of California Press, 1954.

———. *Venezuela*. London: Oxford University Press, 1965.

Liss, Sheldon B. *Diplomacy and Dependency: Venezuela, the United States, and the Americas*. Salisbury, N.C.: Documentary Publications, 1978.

Lombardi, John V. *Venezuelan History: A Comprehensive Working Bibliography*. Boston: G. K. Hall, 1977.

McCann, Frank D., Jr. *The Brazilian-American Alliance, 1937–1945*. Princeton: Princeton University Press, 1973.

Malavé Mata, Héctor. *Formación histórica del antidesarrollo de Venezuela*. Havana: Casa de las Americas, 1974.

Martin, Percy A. *Latin America and the War*. Baltimore: Johns Hopkins University Press, 1925.

Martínez, Aníbal R. *Gumersindo Torres*. Caracas: Presidencia de la República, 1975.

Martz, John D. *Acción Democrática: Evolution of a Modern Political Party in Venezuela*. Princeton: Princeton University Press, 1964.

———., and Myers, David J., eds. *Venezuela: The Democratic Experience*. New York: Praeger, 1977.

Meyer, Lorenzo. *Mexico and the United States in the Oil Controversy, 1917–1942*. Translated by Muriel Vasconcellos. Austin: University of Texas Press, 1977.

Munro, Dana G. *The United States and the Caribbean Republics, 1921–1933*. Princeton: Princeton University Press, 1974.

O'Shaughnessy, Michael. *Venezuelan Oil Fields: Developments to September 1st, 1924*. New York, 1924.

———. *Venezuelan Oil Handbook*. New York: Potter & Co., 1924.

Parker, Phyllis R. *Brazil and the Quiet Intervention, 1964*. Austin: University of Texas Press, 1979.

Parkinson, F. *Latin America, the Cold War, and the World Powers, 1945–1973*. Beverly Hills, Calif.: Sage Publications, 1974.

Pérez Alfonzo, Juan Pablo. *El pentágono petrolero*. Caracas: Ediciones Revista Política, 1967.

———. *Petróleo: Jugo de la tierra*. Caracas: Editorial Arte, 1961.

———. *Petróleo y dependencia*. Caracas: Sintesis Dosmil, 1971.

Peterson, Harold F. *Diplomat of the Americas: A Biography of William I. Buchanan*. Albany: State University of New York Press, 1977.

Petras, James F. *Critical Perspectives on Imperialism and Social Class in the Third World*. New York: Monthly Review Press, 1978.

———. *Politics and Social Structure in Latin America*. New York: Monthly Review Press, 1970.

Picón-Salas, Mariano. *Los días de Cipriano Castro*. Caracas: Ediciones Garrido, 1953.

———, et al. *Venezuela independiente, 1810–1960*. Caracas: Fundación Eugenio Mendoza, 1962.

Plaza A., Eduardo. *La contribución de Venezuela al Panamericanismo, durante el período, 1939–1943*. Caracas: Tipografia Americana, 1945.

Pogue, Joseph E. *Oil in Venezuela*. New York: Chase National Bank, 1949.

Porras, Eloy. *Juan Pablo Pérez Alfonzo: El hombre que sacudió al mundo*. Caracas: Editorial Ateneo de Caracas, 1979.

Powell, John Duncan. *Political Mobilization of the Venezuelan Peasant*. Cambridge: Harvard University Press, 1971.

Rangel, Domingo Alberto. *Los andinos en el poder*. Caracas, 1964.

———. *Gómez: El amo del poder*. Caracas: Vadell Hermanos, 1975.

———. *La revolución de las fantasías*. Caracas: Ediciones Ofidi, 1966.

Rippy, J. Fred. *British Investments in Latin America, 1822–1949*. Minneapolis: University of Minnesota Press, 1959.

The Rockefeller Report on the Americas. Chicago: Quadrangle, 1969.

Rouhani, Fuad. *A History of O.P.E.C.* New York: Praeger, 1971.

Salcedo-Bastardo, J. L. *Historia fundamental de Venezuela*. 7th ed. Caracas: Ediciones de la Biblioteca de la Universidad Central de Venezuela, 1977.

Sampson, Anthony. *The Seven Sisters: The Great Oil Companies and the World They Shaped*. New York: Viking Press, 1975.

Schmitt, Karl M. *Mexico and the United States, 1821–1973: Conflict and Coexistence*. New York: John Wiley, 1974.

Scholes, Walter V. and Marie V. *The Foreign Policies of the Taft Administration*. Columbia: University of Missouri Press, 1970.

Seidel, Robert Neal. *Progressive Pan Americanism: Development and United States Policy toward South America, 1906–1931*. Latin American Studies Program Dissertation Series No. 45, Cornell University. Ithaca: Cornell University Press, 1971.

Spero, Joan Edelman. *The Politics of International Economic Relations.* New York: St. Martin's Press, 1977.

Steigerwalt, Albert K. *The National Association of Manufacturers, 1895–1914.* Ann Arbor: University of Michigan Press, 1964.

Steward, Dick. *Trade and Hemisphere: The Good Neighbor Policy and Reciprocal Trade.* Columbia: University of Missouri Press, 1975.

Szulc, Tad. *Twilight of the Tyrants.* New York: Holt & Co., 1959.

Taylor, Philip B., Jr., ed. *Venezuela: 1969: Analysis of Progress.* Washington: School of Advanced International Studies, Johns Hopkins University, 1971.

———. *The Venezuelan Golpe de Estado of 1958: The Fall of Marcos Pérez Jiménez.* Washington: Institute for the Comparative Study of Political Systems, 1968.

Taylor, Wayne C., and Lindeman, John. *The Creole Petroleum Corporation in Venezuela.* New York: National Planning Association, 1955.

Thurber, Oray E. *The Venezuelan Question: Castro and the Asphalt Trust.* New York, 1907.

Tucker, Robert W. *The Inequality of Nations.* New York: Basic Books, 1977.

Tugwell, Franklin. *The Politics of Oil in Venezuela.* Stanford: Stanford University Press, 1975.

Tulchin, Joseph S. *The Aftermath of War: World War I and U.S. Policy toward Latin America.* New York: New York University Press, 1971.

Vallenilla, Luis. *Oil: The Making of a New Economic Order.* New York: McGraw-Hill Book Co., 1975.

Vallenilla Lanz, Laureano. *Cesarismo democrático.* 4th ed. Caracas: Tipografia Garrido, 1961.

Veloz, Ramón. *Economía y finanzas de Venezuela, desde 1830 hasta 1944.* Caracas, 1945.

Wilkins, Mira. *The Maturing of Multinational Enterprise: American Business Abroad from 1914 to 1970.* Cambridge: Harvard University Press, 1974.

Wilson, Joan Hoff. *American Business and Foreign Policy, 1921–1933.* Lexington: University of Kentucky Press, 1971.

Winkler, Max. *Investments of United States Capital in Latin America.* Boston: World Peace Foundation, 1929.

Wood, Bryce. *The Making of the Good Neighbor Policy.* New York: W. W. Norton & Co., 1961.

Woods, Randall Bennett. *The Roosevelt Foreign Policy Establishment and the "Good Neighbor": The United States and Argentina, 1941–1945.* Lawrence: Regents Press of Kansas, 1979.

Articles

Aikins, James. "The Oil Crisis: This Time the Wolf Is Here." *Foreign Affairs* 51 (April 1973): 462–490.

Baer, Werner. "The Economics of Prebisch and ECLA." In *Latin America: Problems of Economic Development,* edited by Charles T. Nisbet. New York: Free Press, 1969.

Baloyra, Enrique A. "Oil Policies and Budgets in Venezuela, 1938–1968." *Latin American Research Review* 9 (Summer 1974): 28–72.

Bennion, E. G. "Venezuela." In *Economic Problems of Latin America*, edited by Seymour Harris. New York: McGraw-Hill, 1944.

Braden, Spruille. "Latin American Industrialization and Foreign Trade." In *The Industrialization of Latin America*, edited by Lloyd J. Hughlett. New York: McGraw-Hill, 1946; Westport, Conn.: Greenwood Press, 1970.

Burggraaff, Winfield. "Oil and Caribbean Influence: The Role of Venezuela." In *Restless Caribbean: Changing Patterns of International Relations*, edited by Richard Millett. New York: Praeger, 1979.

———. "Venezuelan Regionalism and the Rise of Táchira." *The Americas* 25 (October 1968): 160–173.

"Creole Petroleum: Business Embassy." *Fortune* 39 (February 1949): 91–183.

DeNovo, John A. "The Movement for an Aggressive American Oil Policy Abroad, 1918–1920." *American Historical Review* 61 (July 1956): 854–876.

Ewell, Judith. "The Extradition of Marcos Pérez Jiménez, 1959–1963: Practical Precedent for Enforcement of Administrative Honesty." *Journal of Latin American Studies* 9 (November 1977): 291–313.

Fejes, Fred. "Public Policy in the Venezuelan Broadcasting Industry." *Inter-American Economic Affairs* 32 (Spring 1979): 3–32.

Fenton, P. F. "Diplomatic Relations of the United States and Venezuela, 1880–1915." *Hispanic American Historical Review* 8 (August 1928): 330–356.

Gall, Norman. "The Challenge of Venezuelan Oil." *Foreign Policy* 18 (Spring 1975): 44–67.

Gray, William F. "American Diplomacy in Venezuela, 1835–1860." *Hispanic American Historical Review* 20 (November 1940): 551–574.

Grieb, Kenneth J. "Negotiating a Reciprocal Trade Agreement with an Underdeveloped Country." *Prologue* 5 (Spring 1973): 22–30.

Haines, Gerald K. "Under the Eagle's Wing: The Franklin Roosevelt Administration Forges an American Hemisphere." *Diplomatic History* 1 (Fall 1977): 373–388.

Hausermann, Frederick. "Latin American Oil in War and Peace." *Foreign Affairs* 21 (January 1943): 354–361.

Hendrickson, Embert J. "Roosevelt's Second Venezuelan Controversy." *Hispanic American Historical Review* 50 (August 1970): 482–498.

———. "Root's Watchful Waiting and the Venezuelan Controversy." *The Americas* 23 (October 1966): 115–129.

Herron, Francis. "Venezuela, Its Oil, and the United States." *Hartford Courant*, November 10, 1973, p. 16.

Kaufman, Burton I. "Mideast Multinational Oil , U.S. Foreign Policy, and Antitrust; The 1950s." *Journal of American History* 63 (March 1977): 937–959.

———. "United States Trade and Latin America: The Wilson Years." *Journal of American History* 58 (September 1971): 342–362.

Klein, Herbert. "American Oil Companies in Latin America: The Bolivian Experience." *Inter-American Economic Affairs* 18 (Autumn 1964): 47–72.

Klein, Julius. "Economic Rivalries in Latin America." *Foreign Affairs*, December 15, 1924, pp. 236–243.

Leacock, Ruth. "JFK, Business, and Brazil." *Hispanic American Historical Review* 59 (November 1979): 636–673.

Leoni, Raúl. "View from Caracas." *Foreign Affairs* 43 (July 1965): 635–646.

Lowenthal, Abraham F. "United States Policy toward Latin America: 'Liberal,' 'Radical,' and 'Bureaucratic' Perspectives." *Latin American Research Review* 8 (Spring 1974): 3–25.

Martz, John D. "Venezuela's 'Generation of '28': The Genesis of Political Democracy." *Journal of Inter-American Studies* 6 (January 1964): 17–32.

Martz, Mary Jeanne Reid. "SELA: The Latin American Economic System, 'Ploughing the Seas'?" *Inter-American Economic Affairs* 32 (Spring 1979): 33–64.

Munro, Dana G. "Pan Americanism and the War." *North American Review* 208 (November 1918): 710–721.

"A Nicely Timed Revolution." *Nation*, December 31, 1908, p. 645.

Perez, Louis A. "International Dimensions of Inter-American Relations." *Inter-American Economic Affairs* 27 (Summer 1973): 47–68.

Rabe, Stephen G. "The Elusive Conference: United States Economic Relations with Latin America, 1945–1952." *Diplomatic History* 2 (Summer 1978): 279–294.

———. "Inter-American Military Cooperation, 1944–1951." *World Affairs* 137 (Fall 1974): 132–149.

Rayburn, John C. "Development of Venezuela's Iron-Ore Deposits." *Inter-American Economic Affairs* 6 (Winter 1952): 52–70.

———. "United States Investments in Venezuelan Asphalt." *Inter-American Economic Affairs* 7 (Summer 1953): 20–36.

Rippy, J. Fred, and Hewitt, Clyde E. "Cipriano Castro, 'Man without a Country.'" *American Historical Review* 55 (October 1949): 36–53.

Rosenberg, Emily S. "Anglo-American Economic Rivalry in Brazil during World War I." *Diplomatic History* 2 (Spring 1978): 131–152.

Rostow, W. W. "Guerilla Warfare in Underdeveloped Areas." In *The Viet-Nam Reader*, edited by Marcus Raskin and Bernard Fall. Rev. ed. New York: Vintage Books, 1967.

Tugwell, Franklin. "The Christian Democrats of Venezuela." *Journal of Inter-American Studies* 7 (April 1965): 245–267.

Van Cleve, Jonathan V. "The Latin American Policy of President Kennedy: A Reexamination Case: Peru." *Inter-American Economic Affairs* 30 (Spring 1977): 29–44.

"Venezuela." *Fortune* 19 (March 1939): 74–108.

Welles, Sumner. "Is America Imperialistic?" *Atlantic Monthly* 134 (September 1924): 412–423.
Wharton, Clifton R. "C.B.R. in Venezuela." *Inter-American Economic Affairs* 4 (Winter 1950): 3–15.
Wilkins, Mira. "Multinational Oil Companies in South America in the 1920s." *Business History Review* 48 (Autumn 1974): 414–446.
Wortley, B. A. "The Mexican Oil Dispute, 1938–1946." *Transactions of the Grotius Society* 43 (1959): 15–37.
Zettler, Joseph, and Cutler, Frederick. "United States Direct Investments in Foreign Countries." *Survey of Current Business* 22 (December 1952): 7–11.

Papers and Dissertations

Blendon, Edith Myretta James. "Venezuela and the United States, 1928–1948: The Impact of Venezuelan Nationalism." Ph.D. dissertation. University of Maryland, 1971.
Carl, George Edmund. "British Commercial Interest in Venezuela during the Nineteenth Century." Ph.D. dissertation. Tulane University, 1968.
Carreras, Charles Edward. "United States Economic Penetration of Venezuela and Its Effects on Diplomacy, 1895–1906." Ph.D. dissertation. University of North Carolina, 1971.
Duffy, Edward Gerald. "Politics of Expediency: Diplomatic Relations between the United States and Venezuela during the Juan Vicente Gómez Era." Ph.D. dissertation. Pennsylvania State University, 1969.
Frankel, Benjamin Adam. "Venezuela and the United States, 1810–1888." Ph.D. dissertation. University of California, Berkeley, 1964.
Kreuter, Gretchen. "The Orinoco Company and American-Venezuelan Relations, 1895–1911." Paper. Macalaster College, n.d.
Mohr, Cynthia. "Revolution, Reform, and Counter-Revolution: The United States and Economic Nationalism in Venezula." Ph.D. dissertation. University of Denver, 1975.
Sullivan, William Maurice. "The Rise of Despotism in Venezuela: Cipriano Castro, 1899–1908." Ph.D. dissertation. University of New Mexico, 1974.

Index